CW00720025

To My Supporters,

I would like to dedicate this book to all of those people who have been in my life, but most importantly, who have and continue to support me with my journey and battle with Mental Health. Most of them know who they are, but I'd like to give my thanks to my parents, in particular my mother, who has stood by me no matter what. My lovely dear friend Josie, who sadly is no longer with us and the aids I had in counselling. I'd also like to say thanks to my close friends- Catherine, Georgia, Omar, Sam, Shane, Jake, Greg and many others. Without them, I may not be around today and their support really does mean a great deal to me.

Many thanks.

Matthew

Contents Page

Chapter 1- My Childhood began to crumble

The Childhood stage is supposed to be enjoyable and fun and usually is seen as the best age group by so many youngsters. Live and let live! Being a child brought so much enjoyment and happy memories for me. I enjoyed living every day to the fullest and being so happy made me have great self-esteem and confidence. I would always go out and meet up with friends, we would go swimming or to the cinema for a day out, quality time builds the friendships we have, we at least that's what I thought.

I knew what I wanted to be when I was a child; I wanted to be young and free forever, no worries or any regrets. I never had any regrets as a child, my family loved me, I felt loved and valued as a member of my family, which was quite a big family. I was the oldest grandchildren and had three other cousins at the time, Tom, Marcus and Amy. I never worried about anything when I was young, as mentioned no regrets about missing out of things because I always tried to be involved in them. I had so many great friends and people I knew that I didn't have to worry about point scoring or looking like the best person as they all knew who I was, a young loving boy who had an adventurous and caring side… I had all the enjoyment of going to primary school, getting new toys and games and even seeing the teachers made me happy. Some used to see me as a teacher's top pupil, but between you and me, I just wanted to learn and knowing my teachers, did help me in that sense. I had many lovely teachers, Mrs. Shreeve, Mrs. King, Mr. Gabe and many more, one did however stand out to me, Mrs. Shreeve, she was a tall lady, strict but fair and always had time for our class, we had her in Year Five and she was one of the most outgoing, passionate teachers around. She didn't take nonsense, nor was she a pushover, but my class liked her for that, she was really inspirational. In the whole year with her, we had learnt so much about history, design and technology and the core curriculum subjects, Maths, English and Science. Some of the children from my former class, I do still speak to, most of them do remember this year, I guess it does show that good teaching is a must for helping students reach their full potential. Mrs. King was a nice soul, she had her passions and they did crop up during our lessons, which made them feel more interesting, she showed enthusiasm and we all loved that. She was also outgoing and was very well organized, she was the person to go to for information, she knew a lot about the world and Geography in particular, I recall. And then, there was Mr. Gabe, a merry chap. We had Mr. Gabe in Year 4, he always got the whole class involved, never leaving anybody out, he was quite a tall, thin man, who loved musical instruments, in particular the piano, and he played that every-morning in our school assembly.

Then there was Mrs. Barnard, the school headmistress, she wasn't at St. Mary's school when I first started, but she joined in my second or third year, she was a dainty, well-capable head who had clear visions of what she wanted to do after Mr. Fox left, our old headmaster. She and I had many conversations during my time left at the school, she liked checking on each individual's progress and would welcome those who wanted more from school, and she encouraged me to join the church choir, which I loved. My voice hadn't broken at that time, so I could hit all the high-notes in these hymns, I love it. The choir would visit the old but young, people's homes to spread festive cheer around Christmas, we did this for a few years and it was amazing, seeing their faces lite up as we sung, you cannot beat that feeling. Music does inspire many, mind you now if I sang, my vocal ability is limited, so who knows if their faces would be lit up now? I might get booed.

One final person I'd like to mention here is Mrs. Bateson, she was my teaching assistant for most of my time at primary school, she was a tall white haired lady, who was around 60 when I first met her, but she didn't give up working in the school when she hit retirement, because she loved it so much. She became my rock when growing up, we kept in touch all over the holidays and I often saw her out walking her beloved dog, she loved keeping active and we often wrote each other letters, this happened up until she passed away a few years back, I am always grateful for her presence in my life.

When in Year five with Mrs. Shreeve, I recall I kept asking her what would happen in Year Six, she replied you will learn more about human life and other topics, and then she said

"You'll also be finishing this school and going on to secondary."

Oh no, I remember thinking…things are going to change.

Thereafter, it felt as if my birthday's kept popping up, faster than a revolving door spinning and I didn't stay under ten years old for long, before I knew it I was eleven. Growing up from a child to a teenager was a difficult journey. Some may welcome this change, because more independence is given from parents and some may say, being a teenager is a better life than a child. I did always want to have the double figure age and when I was nine, but when I turned to ten years old, I realized that it wasn't anything special, just another number, just another age. When I got to eleven though, that's when I realized that things were changing quicker than I had thought they could, the years were passing quick and the days didn't stay for long at all. I was then twelve, one year from becoming a teenager… that's when I finished St. Mary's Primary and began secondary school. Since my conversation with Mrs. Shreeve, I feared the change, this was when the regret of becoming another year older hit me, although I couldn't stop nature and make myself stay ten or eleven, I knew at this time, I did not want to move schools and leave all of the people I cared about behind, least of all the school.

Chapter Two- Turning 13

My thirteenth birthday meant I had to change from being a child to a teenager, becoming thirteen should have been an enjoyable experience, the first year in the teens, many happy more teen years to come, but for me, this odd feeling instantly arose and I felt something was changing. I didn't enjoy growing up from a child to a teenager, moving house or my appearance changing. I had never gone through this and the puberty was a huge problem for me. I gained weight and had real bad acne, I recall learning about this in Year Six with Mrs. King, but never did I thought I'd become like this, go through the change from a child to a teenager this quick. I wanted to stay a child, care-free and having fun with close pals.

Sometime after turning twelve and being in the new secondary school, something about me began to change, no longer was I feeling happy, young or care-free. I felt negative about everything.

I had really troubling experiences whilst growing up and I never had anything positive to say or do. I was consistently down in the dumps; I never went out much with my friends or enjoyed life like I had previously. I never saw the good in people, I always saw the bad and thought everyone was trying to control me and change my life. I was scared and fragile. My true life story will show you more about what happened and how my life changed around for the better.

My life as a young teenager was very different to your average teen; my teenage years were very hard and very complicated and frustrating to what they are now as an older teen. I struggled living, regretted everything I did, I had weight problems and even contemplated surgery. I felt like I was shadowing my life and didn't have any enjoyment at all. I never wanted to see friends or family as I didn't enjoy seeing them or socializing either. I was stuck in looking at the 4 walls which appeared to be closing in on me, I felt like a prisoner!

Life was about to get very hard for me. At 12 life was easy, fun and I thought I knew everything, I knew my life, I knew my family and I knew my friends. I was motivated and was always happy; I enjoyed my life and had many friends who I hung around with on a daily basis. I enjoyed making time for my friends, listening to their daily talks, their problems and giving advice where I could, I especially enjoyed hanging out in the parks and having a laugh with them. The Parks were always busy and full of all different people that were there for the same reason as we were, to enjoy themselves and their childhoods. I took pleasure in meeting new people and being out and about. Being out in the big world was fun. It was great to be allowed out and enjoy the company of close friends. I always had listened to my parent's advice not to talk to strangers and to stay with my friends and it was common sense to do this. My friends stuck up for one and other and it was like we had our set friendship groups for the foreseeable future, at least it seemed like that. But life has a funny way of change.

Seeing my friends made me happy, I got bored being at home all the time even though everything I'd wanted was at home. The area I lived in was great, the neighbours had seen me grow up from a little toddler to a child, so they knew me well and they were like a second family as they listened and showed genuine care towards me and my family. It was definitely a nice neighbourhood and an area for people to know each other and talk too. My close friend also lived near me and we often would hang out and chill together. When we were children, we used to hang out most days together, always round one and others house or out and about, the park as mentioned was a popular choice. But we'd also go on short walks and go different places, when allowed to do so.

I really appreciated every day as it came and enjoyed all of my friendships to the maximum. I had so many close friends and each person was different which gave the group a spark. The group and I all got along fine and I enjoyed having close bonds and friendships with these people. I could trust them with my life and hoped they could trust theirs with me. I don't want to go into too much detail, because I am not really in close contact with this group now, but they sure liked having fun, at the time and it gave life a fun perspective from a child's view anyway. The world seemed kind from my view, it doesn't seem that kind now though.

We went on a few years like this, every time we were allowed to go out and have fun, our parents allowed it.

Mum was always welcoming fresh air to a computer game. It would be a computer game on a rainy day, but any other day both my mum and their mums would encourage us to get fresh air, it's good for you was what we always heard.

I wasn't aware that my happiness and enjoyment out of friends would change though. An uncontrollable big problem was arising. Something that I was not expecting or wanted to do happened.

I had a massive argument with my best friend at the time. Once close friends, no longer on speaking terms and the close bond we shared was lost. When that happened to me and the bond was lost, I could not cope with it and to make it worse, it wasn't just any old bond. It was a massive fall out with my closest friend and neighbour. I had known this friend since the age of 3; we went to nursery together, hung out together and had a close friendship. So at the age of 12 this was very, very hard for me. 9 years of knowing someone doesn't just go away, nor do the memories. I couldn't make out at the time why we had fallen out. We all see things in different ways and have different opinion's, that's the point of us being here on planet Earth, having our own valued opinions. Our differences of opinion led our close friendship being taken away and I had been pushed away. So I guess sometimes I should be careful with what I do or say. I blamed myself for years for this fall out, but now looking back at it, it is apparent communication is a two-way thing, he could have made more effort to save our friendship, had he wanted to do so, the sad thing is, even now we hardly see or speak to each other, but one thing I have learnt is never take a friend or family member for granted, their importance in our lives are priceless.

Having thought back to what happened, I recall a row took place. The row was over something so ridiculous and petty. It was mainly jealousy that caused the argument I guess. We had argued about different groups of friends, who liked who best and over who was better at different things like computing, studying, I.Q and being the strongest person. The argument was caused in my view because we spent too much time together. We saw each other in school 5 days a week and after school and then the weekends. Maybe it gets too much seeing the same person over and over again. Some conversations dragged on whilst others were motivated and enthusiastic. When I kept visiting my friend before the argument, it was like the conversations we previously had had were all dried up. Seeing each other all the time wasn't something that we thought would ruin our friendship as that's what friends do hang out and spend time together… but I guess the time changes and people change. It wasn't fair! Being best at everything or being the best friend in a group isn't something that should be judged, if a friend is worthy of being a friend, they will be loyal and help you, never again did I compare myself to my friends, although others may.

Growing up and being so close to my mate made it hard not to fight to get our friendship back on track. We always were together, it was like he was a brother, and it was hard adapting to other people's company. We had always been together in the different groups of friends we had, we had mutual friends and as time went by the groups differed, I never had expected our group to differ as much as it did though. I didn't want the group to fall out; we had so many laughs and good times why should that be destroyed over an opinion?

In my opinion, it was like a much-needed wakeup call had been sent from up above and was out to destroy everything I was and had liked or needed. The group was already shocked at how a simple opinion could break our friendship. They all liked us both but I got the feeling that they would find it hard to talk to each one of us at the same time, they may have thought that they were betraying the other person. I really told my closest friend everything as well as them. Nothing about me was hidden from them as I could trust them with my life and I liked to think they could tell me anything, we would discuss random things and make each other laugh without trying or stupidity. It wasn't like us to fall out; we were always getting along and making sure everything what we did the other one knew about. This was the first big disagreement we had for ages.

After our row, we didn't speak for weeks which eventually turned into months; it was like he really didn't care for his old best mate. I'm not even sure I was his best mate now, looking back at history does make you question what has occurred, perhaps he was just a friend? Close friends do fall out, but not as bad as this. It was so annoying not being able to talk with him or be able to hang back out, but at the same time my pride was just as strong as his was and I didn't talk to him either. I just ignored him and if I did see him, walked by ignoring his existence. I could sense that our childhood friendship was over, it did hurt me. It was an awful feeling, I did have fallouts with people before but when it comes to your best friend it is quite difficult as the bond isn't the same. I had thoughts of turning back time and the thoughts kept cropping into my mind, but I couldn't change what had happened and the argument had been argued so we both thought we were right and ignored the other one, I was very self-centered and he was too. We had many of the same hobbies, traits and enjoyed the same things almost. One big difference we did have though, was that I liked routine and cleanliness, I always would organize my room and try organizing everything else, and he preferred mess and was quite unorganized. Being so much like someone makes you get on like a house on fire and it can be seen as good but it also can cause friction between friends, by being so much alike somebody else. I certainly wouldn't want to have two of me around.

I really was expecting this friendship to last much longer than it did. Knowing him for years and being a childhood friend, had to mean something, right? I wanted it to last long into adult life but I guess I was expecting too much and you never should expect too much. We both knew each other very well, too well sometimes I think. We knew each other's friends and families and we both had seen each other grow up which is very rare, some people do not know their childhood friends in adult life and as he was like the brother I never had, I could relate this. Our families were friends also which made the fall out between us ever more complicated. The best and worst sides of each of us were seen by the other and the fallout destroyed a lot like the memories and times we had, as well as dampened laughs in our friendship group.

Prior to the argument and falling out, we had even thought about our futures and we had made plans together. We discussed girls, cars, gadget's and talked about colleges, universities, careers and many other things in life, like marriage. I wanted to study Law and he wanted to go into the catering industry, I remember that well, because become a lawyer was and is still something that I'd love to do, but getting a foot inside that career is much harder than most think, the degree is needed, then a training contract and with cutbacks, it makes it harder for the ever- increasing popular subject.

I recall asking my mate if he would be my best man when I got married to the love of my life, when I found her. He told me that he would, it was a moment that I could only look forward to and then the argument happened and destroyed the lot! When something like this happens, you either break as person or get stronger, as you see their true colours. I think I broke and it hit my confidence slightly. I was always outspoken and a loud child, everyone can hear loud children and they get heard a lot better than the shy people.

But being loud also has its disadvantages like opening your mouth without thinking, I always said before I thought.

I was in a world of my own, thinking of how I could sort out the falling out. I knew deep down the then 'best' friendship meant something and was worth saving. It then clicked to me what someone had once told me, sometimes in life you have to let go of the things you need most, you cannot have everything or be everything... which is what I needed, everything!

I guess I was very greedy and spoilt as a child, I wanted everything. When a new game came out, I wanted it. When a new film came out I also wanted it and when a new toy came out I wanted it. Mother and father both gave me lots of love and looked after me well, although I didn't always get everything I wanted though as there are some things that money cannot buy such as trust, life, people, happiness and memories. How I see things now though is very different as I understand and know why I can't have everything but as a child the mind is completely different. A child doesn't see that money isn't grown and that money is earned, I always thought the saying money doesn't grow on trees was funny, but now it is completely true. It doesn't grow on trees which makes it hard to get somebody everything they want. It's a challenge to raise a child, it takes real people and parents to do so, so having my parents around, living with me was a real advantage. They were and still are the closest people to me.

My socializing skills were challenged, especially after losing a best friend, as I wasn't speaking to him, ignoring him wherever possible, but yet I wanted to save this friendship. I never had more than one best mate at a time and it got harder and harder as each day passed as we didn't speak nor look at each other or even show sorrow over the argument. And it wasn't a feeling which I enjoyed having at all. The group and I were friends, but there is always a sense of closeness you can have with your best friend, in the sense you can tell them anything and vice-versa. Trust is essential in friendship.

Before I got dropped from the friendship group, I thought I had better leave. It was not fair on them to have to choose between us both, although we both joined the group at the same time. So I decided that I would leave and I stopped hanging out with them and all the times which I enjoyed in the group were ruined like all the other good times and memories I had with my best friend. I didn't want to make any other good memories with someone which I had fell out with. So I didn't hang around with the group as my ex best mate was in that group. The animosity would have been horrible, can you imagine hanging out in a group which is split between you both and the feeling of hate? That's one of the worst feelings I have ever experienced and hate is not nice, it is vile.

He took that group over and tried persuading them that I was the bad person, I wasn't the bad person, an opinion should be heard and he should have listened rather than jumping to conclusions. It wasn't an option to leave that group, in my opinion, I had too! I didn't want to say goodbye to a group that I had been in since primary school, my childhood friends were one of the closest group of friends I have ever had even at secondary school I felt challenged to make new close friends, although I did make a few good friends whilst there.

As I refused to be in the same group as he was, I feared losing the group completely. I walked out on the group after all. All of the trust that my other friends put in me and my actions could be lost as I was a coward for leaving, and didn't fight for what was right. We were not seen in the same group together again. Although I feared losing them, I still hung out with some of them though outside of school and out of the group, but when he was with them, I didn't want to, as I knew it would cause more friction and possibly give the chance for another argument, which I did not want to have. I probably should have, because it became a regret and I hate having them. As if it wasn't hard enough to lose a close friend, I had also lost other friends who had seen the same memories as I had and changes that happened for us. It was made worse as we were still at school together!

After our disagreement it was very hard trying to work out who my true friends were and whether they had sided with him or me. Taking sides is done so often as children, some adults even do it, but I would urge anybody reading not to get involved with problems that do not concern them, life is about you and your family and friends, but never about all of their problems. Carry your own problems, if you have any and try to fix them. Life is easier not carrying the world. It was like another war this time a war against my opinion and views. I didn't want or need another stupid petty slanging match. I hated arguing with people I was fond of, it hurt me as I was a giving person, who valued friendship imperatively. It was like I was being attacked with books, the pain felt awful. The hatred from both of us was shown towards each other and the atmosphere for our class friends must have been awful. I wasn't sure if they would take sides with either one of us or not but I had it in my mind that they would. The friends I hung around with outside of school though, were very supportive and I wanted them to know that I didn't mind them hanging around with me. I just didn't want them to feel like they had too. I know sometimes guilt can trip up and I felt guilty backing down and leaving the group and them behind.

During our argument, when we were forced to speak, horrible and twisted words were passed between each other and some words really shouldn't have been said. We exchanged very hurtful words and the hurtful words cannot always be taken back and I regret exchanging nasty words with a true friend. Another regret to add to my not needed collection. I think that was what made our petty feud and arguing matches prolonged, because instead of me keep feeling like I was to blame, I started to blame his attitude for our row. The understanding of the argument wasn't there. We both still had so much growing hatred towards each other, it was not healthy! I tried being the bigger person by acting really spoilt and I thought that made me look like I was loved more but it didn't. It was made me look like a brat. He acted the same though; I guess we thought we would show off and get more friends to ruin the other person's chance of getting new friends. Again it didn't work. What would work for me? I could admit that it was my entire fault and that my opinion is invalid but I wasn't prepared to do that just yet. I wanted him to apologize to me first, as I had constantly tried to make an effort after the row, even giving up on my other friends in that group, although I did still speak to them. It was a game that was so childish and ridiculously insane looking back and it wasn't very sensible either.

I remember asking my parents for a game for my Playstation 2. I had wanted it for ages and Christmas was so far away that it felt like years, now it comes here ever so quick and passes like the wind, fast! I told some of my friends from the old group outside of school about the game I wanted, then they told the other lot of the old group who still spoke to him, eventually it got back to him and he begged his parents to buy it for him, he didn't even play his Playstation all that much, he was more a snooker person. I couldn't believe it when he came into school with the game I desired, I knew someone had told him that I wanted that game and he only got it so he could boast. The trust wasn't there anymore and I looked in with horror and jealously. Then he decides to play his Playstation on the game I wanted.

Things were already tough between us and it felt like any chance of saving this friendship was well and truly over after this had happened. I tried being polite and acknowledging him, that didn't work, I tried to be nice to the old group but that didn't work either. Nothing I said or did made the situation better and it was hard feeling completely useless like I did. I was sick of trying and that's when more arguing happened. It was so unnecessary looking back, but I had a point to prove at the time, opinions count. Being in the final stages of Primary School made it harder also as we both knew we wouldn't see each other as much anymore and it didn't get any easier, he just didn't see what I saw anymore and it felt like our similarities had gone.

I hated myself for letting this argument control a good friendship and decided enough was enough and all of this arguing wasn't worth it and I tried again to call a truce; I apologized, I showed much of my emotion and admitted my wrongness. All I asked was for him to do the same and for forgiveness too, he had mine but did I have his? It was only over a stupid little thing and if I could drop my pride surely he could do the same? But he refused to apologize and stated that I wasn't his friend anymore as I wasn't capable of being a friend, he then said that I was never his friend, he only used me to get what he wanted. He wanted to be the one who kept the group going around, he didn't really care for them at all. It wasn't right, it was malicious and twisted. All of the years we had known each other and spent time in our group of friends was destroyed, well my friends anyway. I even walked away from them! I knew for sure it wouldn't be long for me to lose the remaining group members I had kept friendships with. I was sure they would take sides with one of us as that is what children do, take sides and it already felt as if, some of them already had done this. It did hurt to think like this and for someone who I thought was my best friend, to tell me that I wasn't "capable of being a friend and was never a friend" really did hurt. I thought I could rely on him and he obviously didn't feel I was a friend anymore, or as he said I never was a friend. I hated this feeling but hated myself more than the feeling of losing a close friend and potentially the group. Was our friendship worth saving or not?

I couldn't believe how twisted he had become, why would he want to boss the group around? They could live their own lives and be much happier without him, they should have behaved how they wanted, not how they were told, but if I said anything to them they probably would have thought I was lying and raving with jealously or worse, hate. I was jealous seeing how close they were but I guess it was all part of his plan to stop me enjoying their company. I wasn't prepared to walk away from the memories and having a few months left in that school, the friendships were worth fighting for, especially the group's friendship more so.

I thought the friendship we once shared was worth saving. No matter how hard I tried though, he never listened to me, he was still blanking me, turning his head the other way or pretending I was a muted voice when I spoke to him. We lived in the same road too, so this made the situation much harder. I lived right near him. He could see what I was doing and I could see what he was doing. Living right near each other made it easy when we wanted to go round each other's house and hang out which was what we used to do on an almost daily basis. It was an epic fail not being able to make up and despite all of my apologies, he wouldn't admit that he was to be partly blamed. Being so self-centered and selfish is never good traits and that's what he was acting like I thought that this friend wasn't truly who I thought he was and after he told me about never liking me or our group of friends, it made my question everything over and over again. I told the group that I didn't want to see any of them again or ever be back in it again and decided that I had to leave them all behind in order to be happy. I ignored some of them eventually like I ignored him. I shouldn't have ditched them all just like that though as it gave those involved and close to me bad feelings and awkwardness towards me. He would also be in control and I expected he would twist everything I said or did anyway. The primary school final year was made increasingly hard due to the argument, hatred and my decision to leave them. They must have thought that I just left out of spite, but I didn't leave because of that. I wasn't go to be in a group where somebody hated me. Hate is a strong word and it is fair to say that I never hate anyone, but I shouldn't have been hated by him.

Chapter 3- The argument continues

The times ahead were about to get tougher for me though… little did I realize how tough they were going to get…

… The row was a big problem but it wasn't the only problem I was facing. The whole primary school class knew about our row and it was devastating. The feeling of judgment came up many of times, I felt I was being judged by them for this row and I am not sure how my old 'best mate' was feeling, but having everybody knowing your business in the class, was not a great scenario. One argument led to a life-changing problem which was losing most of my close childhood friends, the best friend and the group after-all was all I hung around with. I made the problem though as I decided to leave the group and had my opinion as I mentioned. I wasn't ready to forget what I said nor was I ready to try apologizing yet again. I didn't want to hang out with them if they had sided with my foe which most of them did. I did class him as a foe at this stage rather than a close friend, his attitude and words going through my mind constantly, made it hard for me to forget what he had said. He was the one they respected and looked up to now; I couldn't do anything and could only look in. I had the whole taking sides with the other person thing in my head as well as that they were all against me so I went against them instead, well so it seemed. Being a child sometimes makes you realize, when looking back how cruel some people are and only trust those close to you, but how was I supposed to know that at the time. I didn't make a habit of turning friends into foes but it was a larger than life situation and at the age of 12, I didn't always make the right decisions, but then who does.

The argument between us both was still occurring; nothing I said to him could make him admit he was wrong. It was my opinion after all but my opinion was on something he did, so I shouldn't have been the only one to take the blame for this. It got so much for me take and I couldn't get my head around it either. I decided that I had to take a few days away from school as I was physically sick and full of worry. I managed to actually make myself continue being ill and remembered my mother having a few days off work to look after me. She always did care and made that care and love that mothers do. Taking a few days off wasn't my only concern or worry though; I was very worried about the secondary school transition from primary to secondary school which was drawing in! The worrying never stopped for me. My old best mate, the friendship group and I both had decided to go to different secondary schools; in fact all of my class went to the other secondary school except for 4 people! The 4 people that were going to secondary school with me were hardly close friends, I spoke to them don't get me wrong, but there's a difference between knowing people in a class and actually classing them as friends. I didn't hang out with them much or know what they liked or disliked or even what their traits were. It was so frustrating and a real heartbreak, having the thought of this change was like a spiraling staircase, once I thought I was over the worry, it appeared again. I know I shouldn't have minded the thought of them leaving as we had fallen out, but I did and this made me feel so different about the way I saw school and life in general.

Being a child wasn't all fun and games and I had to make more decisions than ever before, infact they were or seemed like they were the biggest decisions of my life at that stage. I didn't feel capable of making a lot of decisions and being right all the time seemed to disagree with me. Parents seem to make most of the decisions when you're young, they know best or that's how the saying goes. My parents always did know best, what was best for me, who was genuine as a friend and supported me in any way possible. Despite not getting along with some of those in the class, I knew that if our friendships were to ever get back on track that this needed to happen before we left primary school and there wasn't long left either. Otherwise things wouldn't be the same as previously, I started St. Mary's Primary school at the age of 3, so now at the age of 12, almost 13, it was hard to think that I would never speak to them again and everything we shared would be gone forever. We wouldn't see each other every day, the 5 days a week in school would be gone and this was a scary thought. I hardly had any days off school, well primary school anyway, because I enjoyed it so much throughout, this dampener was infact the only negative of the whole primary school experience, although I do see now that an increase of bullying is occurring and must be stopped, so other young students do enjoy the primary and secondary school experience.

I had to get the friendships back on track and now. I wouldn't have had the chance otherwise and would regret not doing so or even attempting with them. More regrets would be unbearable to deal with, they make people seem different and out of character. Even if none of my friends had been speaking to me since I left the group and even if my best friend still wasn't talking to me, I had to get heard, I had to at least try. I helped many in primary school and even now, I try with everyone I meet, always trying to help them reach their potential and offering support along the way, sometimes I try a little hard, but a trier is what I am. So I tried and tried but wasn't being heard and despite this I wanted to continue trying to mend our friendships before we left for secondary school permanently but it was harder than words could describe. I went and asked a close teaching friend of mine, Josie was her name. I asked her for her advice and she looked at me quite concerned. She had a lovely character and was somebody that I thought I could trust. I did speak to my parents about this, but their advice was to keep trying and the determination of this was draining me. Josie asked what had been going on so I explained and she seemed to have the solution, if I wasn't in the wrong then I should keep fighting till the end.

"Why should you not stand up for something if you believe it, Matt?" She questioned?

"I agree Mrs. Bateson" I replied, "but how can I ever give up on people I once classed as a big part of my life?" I remember questioning.

Then, I told her I was partly to blame as it takes two to argue and she agreed but encouraged me to continue to try. She told me that if there is something you want so bad you need to show you want it before it goes,

"Life is too short Matt" she exclaimed. "Never have regrets and never feel you cannot do anything, because you are an incredible young man." It was a sentimental moment and I could have bawled out.

I could see her point clearly and I decided I wasn't going to accept being rejected… I would try again, if the other times hadn't worked, then I mustn't give up, she was always around for me in friendship and through support.

Petty arguments were not even worth having anymore and I couldn't believe that I wasn't being listened too. I just wanted to fix my relationships with my friends before it was too late and they forgot about me or thought I didn't try, after all I encourage/d others to try, so I should continue trying. Being forgotten about was my biggest fear. It was already hard to accept change and to grow up and the thought about going into secondary school was scary, it felt as if everything I had known was being taken away from me. My life felt unpredictable.

After much time trying to get their forgiveness after leaving the group, I was finally forgiven and my pleas were finally heard. But this came at a price and I was forgiven but not welcomed back into the group; I was still the outsider looking in. I knew that he had something to do with their decision. This wasn't like them to be so hasty or unpleasant. The feeling of anxiety came along, I had never felt it before, but at this age of just under 13, I did feel anxious. Having said that, I did feel anxious about the end of year exams we were having, but the feeling was not as bad as this one I experienced here. I was sure they had a plan and I needed to think of a better one. I didn't know their plan though or even know if they had a plan, I assumed they did, I recall lying awake at night worried over-thinking what was going to happen and what was going on then. They were so harsh not allowing me back into my friendship group, nobody owned that group I thought and we were in all that group because of our friendship after all, so was their forgiveness genuine? To make things worse for them, after he got what he wanted and destroyed their trust for me, my old best friend left their group. He didn't really like them; he used them and played the sympathy card as I was in the wrong supposedly. He did tell me, as mentioned, that he wasn't really genuine but he had to always try to have the last laugh or joke, it made him get off I think, seeing other people suffer, who actually cared for him, but then I couldn't say anything, because I thought they'd see me as the liar. He decided to go with another group of' mates' that he had before us. It was very complicated and I didn't know why they wouldn't allow me to rejoin the group as we had been through ups and downs together and had memories together. Thinking of it now though, I can see it was due to misjudgment, something that happens all too often in this life. It was like a piece of me had gone and went away; as soon as they told me that they didn't want me back in their group I was devastated. It was not fair on me, I only left the group because of my differences with him and it wasn't my entire fault over the argument, I was only partly to blame. I eventually decided their price wasn't worth paying as it still would hurt me not being in the group for the final weeks. I was annoyed so much that I told them to forget it and I didn't want to join in their group or need their forgiveness. But of course, I did want this, my words may have said this, but surely they could see my actions. I guess I should have explained the reasons behind my decisions? There's nothing worse than feeling unwanted, I've known this feeling because of this, but it hurts to walk away from something that is loved and cared deeply about…that's the part that most people forget who do walk away.

The group obviously didn't want me in it and it was like I was talking to a brick wall which was hundreds of feet tall, with them and I certainly was not getting anywhere. If I could apologize and be willing to forget and forgive, then why couldn't they? Being children at the time made it challenging as we all had very wild, adventurous minds and changed friends by the day but best friends always stay. Peer pressure was also another factor, if the group didn't want someone in it, even if one or two people did, then that person was not joining the group. They didn't want to know me but did tolerate me. It was rather difficult to tolerate their toleration of me however as I wanted our friendships back, but wasn't getting them back. All my attempts and efforts were not making a difference and I even tried to buy the good times and friendship back by offering my games but that didn't work, nothing worked. How daft was I? I'd never try to buy people nowadays, they can either be my friend or walk on by, and you get the wrong people in your life if they need incentives to stay. I knew Josie would say the same, she had a similar opinion as me when it came to friends, but never did I see her, not once, have a bad word to say about anyone, she saw everyone as equals, something so many fail to do, due to past experiences like this or worse, I guess.

All of this working, to try and get back into the group and even trying to sort his and my differences out were not working and I was still trying to get forgiveness from him which made the situation worse. Why did I even waste all of this time? Time is so precious and certainly should not be wasted. My attempts to be accepted back in the group kept failing and I just gave up and wasn't able to succeed with them. Josie must have seen us from the school window, the school was a flat school and had no stairs, the playground wasn't all that big either, but big enough for the small school, she suddenly came down and asked me what had happened with her sympathetic ear, I burst into tears and told her that I couldn't win them back. I told her about him leaving the group and he must have brainwashed them and forced them against me, I also told her about them 'forgiving' me, yet it felt like it wasn't genuine. She comforted me and gave me some more encouraging words.

"The winner is the person who never quits Matt and allows defeat to beat him, he gets up and continues fighting for what is right."

She knew that I wanted my friends back and she was the light at the end of the dark tunnel for me. She was very wise and had much wisdom, I wish I had that. So I thought I must try… people say with age comes wisdom, but I rather see it from experience comes wisdom, you can grow old and have little wisdom, if you dare not explore yourself and the world around you. Change occurs with exploration, wisdom comes with that, not age.

My attempts took much, much longer than I thought they would. The days didn't stay still for me and weeks went by quicker than I thought they could. The secondary school permanent transition was coming up. I wanted to stay on at primary school with the people I grew up with, I loved the primary school and everyone knew each other and had childhood friendships like I had and even the teachers were kind and you could speak to them about anything, remember I mentioned Mrs. Shreeve, Mrs. King and Mr. Gabe, well they were just a few to name that offered great kindness and amazing teaching. Mrs. Turner also was a great support throughout my time in St. Mary's as was Mrs. Peacock. Mrs. Turner was the head of the kitchen, always cooking lovely meals and saving sneaky extras for me. I loved that and her puddings especially, they were delicious. But that wasn't why she was supportive, she listened and guided many like myself with the whole school experience and went above and beyond her pay grade. I keep in touch with her now and she certainly is one in a million. Mrs. Peacock was the school caretaker and the time, she had lovely long red hair, and always smiled. She was a happy soul, I am sure she is now, I chat to her occasionally and recall her pleasantness around the whole school and to everybody she met. I had great friends, I had guidance in Josie too, she was extremely helpful to me and she supported me with everything and gave me advice when I needed it, the examples I've mentioned are just a few, I could write a book on just Josie. She and I had a great friendship and we always spoke to each other and sent each other birthday and Christmas cards. I was worried that I wouldn't see her again either but we kept in contact as usual which was good as I didn't want her to think I had forgotten as she was my support, infact she only lived around the corner from me, so I did pop over to see her often, not as much as I could have though, but we did keep in touch through letter, quite often actually.

After having so much enjoyment out of primary school and knowing almost everyone in the school it was very hard to say goodbye and let go. What if I didn't see them again or they didn't remember me? I was left worried about the transition for weeks and the bad thing about it was it was a long holiday before I went to my new school, just over six weeks, the same as the summer holidays now.

Days went past and eventually primary school was almost done! I still didn't have the genuine 'forgiveness' from the group and nor was it likely to come.

The last day at my first and best school ever, had now approached and was here. I was so upset that I showed my feelings and cried. I wanted everyone to know how much I would miss them and I didn't want to change or go to the secondary school. Mrs. King was our teacher at the time, she made the day special. She got loads of treats in for us, even buying us fish and chips at the Chill-Out Club! She ran this club once a week for Year Six pupils and many went, it was the final club we had as it was our last day. I thought that my old friends would have forgiven me by now for leaving them. it added to the hurt and pain I was feeling about leaving, as well as losing my friends, the teachers would also be missed they taught us so much which we often take for advantage, even the basic how to write, how to spell and how to count, they count towards a great deal of adult life, yet are often underappreciated.

Josie again came down and saw me, she told me how much she would miss us all and that if I allowed defeat to win then I would be the loser. She could read me like a book I guess. I thanked her for all of her help and support. She said I had nothing to thank her for and it was her honour to know me. Funnily enough she retired that year, she was like a guardian angel through the whole of St. Mary's school to me, but yet she didn't need huge thanks and appreciation, she just was happy to have helped. How rare is this?

Meeting somebody who cares, support and guides you, yet wants nothing in return? She was one of my best friends and somebody who has touched my heart, I gave her a smile and knew today was the day I needed to fix my friendships for good. The final chance and attempt must work.

Chapter 4- Leaving St' Mary's School Behind

Before I attempted to get their forgiveness though their genuine 'forgiveness', I went around taking photographs of everyone I knew, all the different year-groups and the different teachers all striked a pose and flash that was that. I don't recall what happened to these photographs, but after getting them developed I remember sobbing away, it become reality that I had left the school, something I had sleepless nights dreading.

We had a leavers assembly on the last day, everybody was emotional and was waved off by the younger students within the school, the year-group , Year Six, then went down to the church and was given a signed bible from Mrs. Barnard, Head-Teacher, and shown our leaver's DVD. I remember watching the DVD and remembering all the good memories and times I had within the school. We had a trip to the Isle of Wight in Year Six, and Mrs. King had taken photographs of us on the trip, the 4 day visit was great, it was my first holiday away from my parents and with friends, we had a great time. Funnily enough the trip was called PGL, which apparently stood for- Parents Get Lost. It did and still does make me chuckle. Below are a few pictures of me at PGL.

Back in the church, I turned my head and saw my old friends, I smiled at them and they smiled back. It was like we had made up, they must have seen what I had seen, fear. I never revealed this to them whilst we were not getting along, but they must have somehow saw it. Fear has a funny way of appearing, even when we least expect it or want it to. Josie was also in the church, she was sitting on one of the long, wooden benches, there along with the other teachers. She smiled up at me and I had the strength of friendship and understanding. It was very emotional getting the bible as I had seen the other previous year 6's get their bibles before I did. Having the negativity about secondary school didn't really help me as I wasn't ready. Each of our names was called and we collected the bibles and shook the vicar's hand, the vicar at the time I believe was Father Arguile, his daughter was in our class during the time. He had been in the church since we were at first started school, so he had knowledge of who we were, especially the ones in choir. It made it even harder to see leaving as a good option; it wasn't good it took everything I ever knew away, my daily routine would never be the same after this day, the people in my life would change and I did not want this at the time.

After a few hours at the church, we all went back to primary school and I remember packing the stuff up in the church and driving back to school with a few other student councilors and the head-teacher- Mary.

Mrs. Barnard was so proud of us all and she told us this on the way back to the school, she had seen us all grow up from little children to almost teenagers. Seeing your head teacher like that makes you think, wow I have been noticed here and know everyone and everyone knows me. Although she wasn't at the school when we all first started, she was the longest head teacher we had. I believe she worked for almost eleven years in St. Mary's school and my, did she make a positive change! We must have been a good year-group. She was extremely sad, but she didn't admit she was, but I could sense these feelings. I had experienced sadness and worry and fear. It was nice to have a final discussion with her as well, she was caring and admitted she would miss us which meant a lot as she knew us all and everything about us. She was like a piece of the family, like the other staff I have mentioned in my book.

When we got back to primary school, I went round the playground which I had been in for 9 years, I went round looking around and saw the places where things happened and had changed. The things I noticed was the playing field, the park and the Frog waste bin. It is strange what you see when you get emotional and full of worry, but what I saw was what I had been seeing for that duration of time in school. Whilst walking around the playground I was getting people to sign my bible. Some of their comments were very inspirational and moving and most of them showed genuine, true feeling and said they would miss me. Tears started pouring out of my eyes again and I spotted my old group of friends just standing on the playground. I couldn't see my ex best friend with them so I ran up-to them and asked them to sign it. I didn't expect them to sign it, yet alone talk to me after all what had happened but they all signed my bible and told me to stay with them as they wanted one final day with me and make amends. Things changed and they must have known I needed them. It was so great to eventually make up with them, I felt complete, even if I did not have a best friend on side. Josie's words on encouragement and commitment towards me paid off and I was able to be with my friends in primary school for one last time on the last day, I only wish I didn't leave them now, but the awkward atmosphere would have made me ever more anxious. I told them how much I regretted what had happened and was glad to be back with them. They told me they were glad to have me back as the gloom of the group was controlling their thoughts and keeping them against me. I remember a sadistic thought I had, it was that the traitor had a plan and then a possible revenge plan cropped into my mind but why should I be so twisted and lower myself to that level again? I let it drop. I let the positive moment outshine the fear of negativity. We all signed each other's bible and then I remember questioning them about their feelings over the transition and leaving everything we were behind. Oddly enough they felt exactly like I did, they were worrying over the move and about the future. None of us wanted to grow up and move away from each other. We all would have stayed in the primary school if we had the chance, we probably still would be there at the age of 18, but that simply isn't possible. Life changes occur so nothing can stay the same and even we cannot stop change. My perseverance paid and being a 'trier' sure did help me. I felt that this last day was one of the best I had in weeks, even if it was that, the last day!

The school whistle went and it was time for our next school assembly. We had one final assembly with Mrs. Barnard where she gave a speech to us all, she wished us well for our futures and told us all how much she liked seeing us all achieve, grow up, become different characters and that she would miss us all. She then even burst into tears which was strange to see as most teachers don't like students or at least it seemed like that, in some schools. But not our school.

Every child was pushed to reach their goals and guided. I sensed that she was upset in the car journey back from church earlier that day, so my sense was correct, in a way I was glad another person was crying instead of me but I did feel bad for her. She was lovely and I went up with the other student councilors and gave her the biggest bouquet of flowers, I had ever seen, to thank her for everything she had done for us. Josie was also retiring at the same time as us leaving, so we had some flowers for her and a box of chocolate. Luckily they both liked flowers! She was extremely grateful as was Mrs. Barnard. She then continued on to announcing that she would like us all to stay in touch with the school and let them know of our progressions and ambitions.

Josie was happy I think, she had worked so hard for many of years and went above and beyond what she did in her work. I thanked her again and she said that I didn't have to thank her, she was happy to help. She was so genuine. Someone being this kind to me, other than family, was appreciated more than words can describe and she was amazing. I guess without her help, encouragement and support I wouldn't have stood up for what was right and that was proving a point, even though some may read this and think I should have put my opinion aside, it's very hard to admit you're wrong, if you feel like you're right. I hope you can relate to that.

After Mary's speech, the class was allowed to go to the chill-out club, where Mrs. King was happily standing. We played games, Ping-Pong, table tennis, even snooker I think, which I must admit I am not that great at! I can never hold the cue correctly. We then watched films and ate fish and chips which the school must have brought for us, but Mrs. King got the appreciation as it was her that spent time organizing and running this club. It was nice to spend one final longer than a school-day, day together as a class, everyone got on for this event, even if we did have differences, and we could talk like friends again. I may have lost the best friend I ever had, but I had other friends and they had forgiven and understood why I left. We all then played party games which Mary had organized as a surprise. Janet (Mrs. King) and Julian (Mr. Gabe) came down with more surprises: a goody bag for each of us and good news, we could continue playing on the snooker table and table tennis game, we all enjoyed the chill out club and great fun times we had in the club.

Time goes much quicker when you are having fun though and the time was ticking and it wasn't long until I waved goodbye to everybody, it was around 6pm in July, so the days were still light, it would have felt absolutely doom if it had been a winter night, all the darkness. It was the hardest thing ever to do as I had become attached to the school and my friends. My perseverance even got my friendship group back in the end.

I hugged all of them, I exclaimed that we all must stay in contact and stay close friends. After months of feuding we had finally made up and this was great although we wouldn't see each other all the time, we could at least still be friends. It wasn't all bad I guess, as it would have been if I hadn't made up with them, I appreciated their forgiveness, even now they don't know why I left, but can probably understand better being in their twenties. My 'best friend' had left without saying goodbye to any of us. I was sure that when that was mentioned the others were somewhat annoyed and bothered by it, but had more important things to think about. We all told each other a special memory of what we had of one and another and one we will have forever. It was a special moment. Infact it felt like a heart-touching moment. After our heart to hearts and moving on with accepting change, we got our bibles and went around saying goodbye to the teachers for good. We all went and said our final goodbyes, we said goodbye to Josie first and she was sad to see us go, but it was also her last day which must have been hard for her also, I was sure she had been in that school for many years, she loved her job. She smiled at me and whispered:

"See Matt, I told you that your strengths and abilities will pay" she was right. I guess we all allow our feelings to sometimes take us over or they hide deep within us and we struggle to reveal them. She knew me inside out after much time and she saw what I couldn't. I was extremely lucky to have known her

She was leaving before we left, we all waved her off. It again was emotional, but I didn't cry, I don't think my tear ducts had recovered after all of the crying previously. I struggle now to cry, sometimes it would be easier than bottle things up, but its difference being 20 than 12. I was sure that I would stay in touch with her as she had seen me grow and become a stronger child. She helped, encouraged and explained, I had the sense of care from Josie .I was encouraged to stand up for what is right and found my abilities. We then went for a photograph with Mrs. Barnard who was glad to see us one final time, she had her photo with us and we all thanked and congratulated her on the success of the school and for her time for us all and for everything she did for us. A few more tears could have come out and as we walked out of the reception, I knew this was the final time I would take that walk. It was like I was walking out on my old life into a new one, a change that was not wanted.

After all of us walked out of the school, we all departed and went our separate ways home. I got home first as I lived the closest. I lived in a lovely flat, some saw it as plain, but is wasn't. It was my home and one I was raised in, I had many good times in it with friends from the school, neighbours and family members. It wasn't a time that I wanted as I couldn't go back into the past or rewind time, and I knew that primary school was well and truly finished. It wasn't a mistake that this was happening, it was a change and that happens. It had to happen. I shouldn't have feared change yet I did and the transition didn't happen overnight either. The summer holidays had to come first, they were ever so long that year. Well that was one vital chapter in my life done and over with. An important part of me disappeared when I left that school, but it was inevitable that that had happened. And I know that it isn't just me that worries about the change, many of you do. If only it is made easier for the transaction to happen, without the apprehension. After all those fears about never being able to make up with my friends, I was hurt more by having to say goodbye to them. I had been having sleepless nights fearing that they would not forgive me, but now having their 'genuine' forgiveness, I feared losing them. Fair enough I could probably see them during the long summer break or after secondary school finishes, but I wouldn't be seeing them constantly, on a daily basis, or seeing them mature into their teenage years. I made up with them to leave them behind; it wasn't making any sense to me why this had to happen and I remember the feeling of loneliness cropped up and made me feel ever more anxious.

The summer holidays had approached and usually I enjoyed this time. 6 long weeks away from school and being able to do whatever I wanted or enjoyed most, it was amazing to have this long break away from school, but this 6 weeks holiday was completely different to any others as it was like 6 weeks from hell as I didn't have anything else on my mind except this big fear. I don't fear much but this fear felt like it was destroying my happiness and taking over my mind and life as I didn't want to move on, I didn't feel ready.

The first day of the holidays seemed awful, I remember I slept in and my alarm didn't go off, the batteries had run out. I had that alarm for 5 years and never had to change the batteries and it did seem peculiar that it had done now, on the first day after finishing primary school. I woke up thinking that I had to get dressed for school and was rushing to get ready as I was late. I got my uniform on and went to go to the school but when I got to the school gates they were closed and chained. It all came back to me that I had left and that was no longer my school! It did take some time to get used to this change, I guess I wanted to live in the past again.

I walked back to my home, which was only around the corner from the school, so not that far, and I sat in front of the television, there was nothing to do and all that was on the television was debate shows so I switched them off. I was not interested in hearing about politics as a child, I found it all too dry, but now I am very open for a political debate. I enjoy them! Having turned the T.V off, I sat in peace and quiet and was full of resentment towards the idea of going to another school, I didn't want another chapter, not just yet. I think my parents knew I was worrying more, what made things worse was that I wasn't talking to many people, I did feel different for sure.

I didn't want to change; I didn't want to go to another school. I had been in my first school for more than 11 years and saying goodbye the day before was the hardest part ever! I kept telling my family this and they seemed unable to grasp why I didn't want this to happen. They must have all gone through this stage as children, so knew how it felt, but everybody is different, everybody has different feelings and emotion, so perhaps they didn't feel the same as I did? Despite my attempts to stop them from sending me to secondary school, when it began, it seemed as if my family's encouragement made me see why I had to go. I then realized that I finally had to accept moving on and growing up, hearing their comments and opinions really did show me that another path and chapter had to be walked.

Moving on was very hard for me though as I didn't know how to, I just wanted to stay at primary school, I keep saying this but it really was how I felt; I wasn't ready for change or leaving my old school and friends behind but I had too and my thoughts wouldn't make it all alright.

Once I eventually saw that I had to leave primary school behind, (I had already left and hadn't thought of moving on), I had to shut the door and stop wishing that I could go back and see everyone again. I didn't see anyone much during the long holiday break and it was awful. I believe I just stayed at home and did different things with my parents, my father was working from 7am- 6pm, so he was hardly home, but my mother booked holiday for some of the long break, so I spent most of my time with her and visiting family. My old friend whom I argued with still wasn't speaking to me, I didn't have anyone to confide in except family and it wasn't the same as they wasn't going through this transition at the same time as me, I stayed awake most nights thinking about what to do, whether I should be going to this secondary school or leave for the other one where everyone I knew was going too? But it was too late to even consider this, because when you reach the final stage in primary school, you along with your parents have to decide on what secondary school you attend and I had chosen mine already.

It was terrifying, not knowing what to do or what was going to happen at secondary school and the fact that only 4 members of my old class were going to the same secondary school as me was also daunting.

I even planned to run away from my family home, it was an idea that came into my head a few times. Perhaps if I could go somewhere else, somewhere where nobody else knew me and couldn't tell me what I can and cannot do then everything would be alright. It seemed a good idea at the time and I could really see myself being able to cope being alone. After all it seemed that none of my friends cared for me no longer as none of them bothered knocking for me or even calling me to see if I wanted to hang out. I was still annoyed with my best friend for not even talking to me. He could of at least said the final goodbye if he didn't want my friendship but nothing. I had packed a small suitcase despite not having any money or knowledge of where I was going to run too. My ideas always seemed right back in those days and I always listened to number 1, not anyone else. My little red suitcase, was only about 60 centimeters high and not even 20 centimeters wide, it certainly did not fit a lot in. I didn't run away, I may have thought about it, but thinking back I see I would have never coped without my family or home.

After that happened, I went and visited Josie and told her that I couldn't cope with not being able to control my life how I wanted it to be, she told me that I shouldn't worry as all will be fine. I think she thought I was being over-sensitive or worried for nothing but I couldn't see that at the time. I thought she had no compassion towards the matter and went home. How wrong I was, she had much compassion. My suitcase was under my bed and I grabbed it that's when the telephone rang and my mother answered it. It must have been Josie, I think she told my mother about my visit and that's when my plan to run away was ruined. I had been caught with the little red suitcase in my hand and was told I wasn't running away from home or secondary school, I was going through with it. I didn't want too and that was clearly shown and expressed.

Everything was getting on top of me, my shoulders felt weighed down by bricks but they were my problems. These problems and nerves stayed with me for the rest of the summer holidays, I had never experienced so much worry before, it wasn't even as bad as the time I left the group, I could cope with that in some ways, but this 'transition' I couldn't cope with. I never accepted a great deal of help so nobody offered it to me, I was very independent and could do most things alone. I was taught very well by my parents, how to think by myself, do chores alone and how to do daily tasks. It is great being taught so young, because now I am able to do most chores, except the ironing, I never attempted that. After the conversation with mother about why I had the suitcase and my plan to hide from the transition stage, I was grounded for 'playing up' and it annoyed me so much. I liked to go out and about even if I wasn't doing much of this as my friends were not knocking for me. As I was grounded it made it hard for me to think about anything else, my thoughts were well and truly on fear and the fear was taking my happiness away and I wasn't talking much either, I bottled this feeling up for some weeks after, as I thought I could cope after telling mother and Josie about my worry.

The phone then rung again, I feared that it was somebody else I had told about my fear too, I did tell my Grandmother, Sandra about it, it annoyed me that I told a few people with my entire trust and they told my mother about it. Trust is there for people to talk to others in confidence about little or big issues but I guess my fear had to be shared, just in case I didn't tell mother. My Grandmother was always passing on her opinion and views, she continues to do so to this day, but that's good, because it shows she cares for family, my family. The phone call however wasn't from her or anyone else telling my mother of the fear I had, she already knew of, no it wasn't anything bad, it was my Aunt Emma and Uncle Tim.

Chapter 5, The Temporary Getaway

My Aunt and Uncle had a surprise for me, a very nice surprise indeed. They did a very kind thing for me, without me having any idea until a few weeks before, and they offered to take me abroad with them. It felt like this offer couldn't have come at a better time and they took me abroad with them. I could get out of this lonely phase and away from the constant vicious circle in my brain, which was putting me down and fearful, even now I dislike change slightly, it's something I have never felt good at, even if it is needed, if my life is good at that stage, I wouldn't welcome change. I was very, very happy with this invite and my cousin, Marcus, was also so happy that I was going with them. I also have and always will class him as the brother I never had. He often was round ours and he grew up with me, so I was really looking forward to going abroad with them, in particular my cousin, there is a five year age gap between us, he was born in 1998 and me 1993; it was the first time that I ever went abroad I never even had a passport, because my parents were happy to visit England's holidaying sites, such as: Cornwall, Great. Yarmouth, Hembsy, Billing, Norfolk, Cromer and Caister on Sea. Although there is nothing wrong with these places, going abroad was different altogether.

After getting the good news, I felt happier, something I could focus on, so I decided that I would go and get some new clothes and summer things like shades, sun lotion and flip flops. It got my attention focused on something else and was very positive rather than negative. More optimistic than pessimistic. I was so looking forward to this holiday, it was something I thought would take my mind off of the transition, although I was worried about not seeing my parents for a fortnight, I was assured that they would still be there when I returned and they wouldn't change. It was just the thought of being away from two of the closest people to me that worried me, but I needn't have worried that much, they didn't move, luckily!

The day to leave for the airport came around quickly, not long had I finished packing, it felt as if it was time to depart and get in the car to go. Enjoy yourself I was told, so I tried my best to leave the worrying behind and looked forward to the journey there, knowing what was on the other side. I was worried about being on a plane for the first time though but it was a real great experience, little or big things I worry about, little did I know I was to be diagnosed with anxiety later on.

After the journey to the airport, Gatwick airport I believe, our cases were taken out of the car by father and Tim, Emma, Marcus and I was waved off by him. My other Aunt and Uncle, Emma and Mark, also were coming along on the holiday, which was a surprise to me, as I was told at the airport. All of this did feel surreal at the time, I had been worrying for so long about the transition and then this happened.

After we got through security and the gates, we eventually boarded the plane and got to our destination. We went to Fuerteventura in the Canary Islands. The scenery was amazing, hills and mountains in the distance with sand clouds and hot sun. It was like my dream, the sea water was pure blue, not like any I had seen in England. The hours we spent on the plane were sure worth it, despite the journey feeling like 10 hours! We then got a bus to the hotel and the hotel was lovely, it was within walking distance from the beach and there were even 7 large swimming pools with bars onsite! Obviously I could only have soft drinks but I didn't mind that as I was more looking forward to being happy and exploring a different place, alcohol didn't appeal to me anyway, I had heard so many bad stories of what it did to the body, I was disgusted. Upstairs in the hotel there was a lovely restaurant which had all different types of food in. Continental food was served up all day and the food was fabulous, some of the food I had there was unusual but tasty and never tried before. I couldn't moan at being away from home as this was like paradise. Being all inclusive was great, free ice-cream and food all-day! I somehow managed to focus my mind on something else other than the school transition and sent this worry to the back of my mind for the first time since knowing! It was great!! I was really happy and enjoying the sun, sand and sea, it was such a great beach and I met a few more friends along with my cousin. It really did help me mentally, this holiday. I didn't expect this, but Marcus had a difficult question for me and I remember his question well and it was so sad, his question was:

"Why have you been so upset Matt? It isn't like you to back down and cry like you have been."

I couldn't answer this; it was too upsetting and a fragile matter that I tried to hide from. I couldn't and wouldn't frighten him about the transition I was scared of, not for the world. I didn't want him to know about my fear and I didn't want him to have the same fears as I once did when his turn many years down the line to transit happened. My confidence may have been sidelined but I thought it was easier not admitting my fears to him, although they worried me more than any opinions, so I replied:

"It's just my hormones mate; I really struggle with them at times."

What else was I supposed to tell him at the time? He was only around 7 years old. I just put up a front and told a lie. But this lie was a good lie, it hid my emotion and I was never one who showed my emotion much, especially not to anybody who looked up at me., unless I had bottled something up for so long, that's when I lost control, because all of my emotions burst out. I had little control over this, when it happened. I wanted what was best for my cousin and his future, so I ignored my worries and this felt like the right thing to do. When he got to the transition stage, he may have seen it as a something to look forward too as we are all different and see things in different ways. It was a real emotional chat with my cousin, but my fear was better hidden from him, he probably wouldn't have understood anyway.

After our heartfelt conversation, Aunt and Uncle wanted us to go and explore Fuerteventura, and wow, was it a lovely place, I recommend it highly. So we both left the hotel and went over to Costa Calma, which was 12 miles from our hotel. We decided we would catch a bus there and surprisingly walk back the whole 12 miles, across the long beach which had hot white sand and pure blue seas. I really felt that I had a close companion as well as a family member in my cousin, I could speak to him and he would understand or made out he understood, of course I didn't tell him everything though. The journey to Costa Calma was great, everything seemed to amaze me. The lifestyle was completely different and there were other children that looked Spanish running down the beaches laughing and playing, do they even go to school full time in Spain? I remember questioning. When we got into Costa Calma that was when the holiday got even better. It was over 40 degrees there and the sea was light blue, all of the sea seemed that lovely colour, with a sand bank in the middle of it! It was a picture moment. Sometimes being on holidays allows you to think about your life, a holiday is like a rest that everyone needs but not everyone has. I was becoming used to being out in the sun without a fear in the world. It certainly did help me think about my life, what has happened so far and what I wanted to happen in my life. I remember the thoughts of becoming a lawyer keep popping up whilst on holiday and I sure still have this ambition now, maybe I can achieve this one day.

It felt as if I was becoming a different child, I had been thinking about things more clearly since being on holiday, I realized that not getting everything I wanted was a bad thing, it is just that I couldn't have everything I wanted. The value of money was being shown to me and I needed to accept this. On holiday I had my own money to spend but I didn't spend most of it, I only really brought souvenirs for my family and hardly anything for me! Understanding the money as a child is important for adult life and again this was something that I had underappreciated, until I had my own money and had to budget for myself.

That's when I had another thought. I had a strong, supportive family, who cared deeply about me so much and it really did help. I know some people who I support through BeatDepressionTogether, who don't have their families to help them and thinking about what I have rather than what I do not have, made me stronger. Value everything whilst you still have it, never underestimate anything in your life, because it could be taken away as quickly as a pin dropping to the ground. I accepted that not everyone in life will like you, everyone has diverse egos, characters, opinions or judgment, it wasn't me personally that was hated, but rather just not liked by the whole world, and not everyone forgives you either, although the only forgiveness I needed was from my friendship group, what I had, and finally not everyone in life will be pleasant. Life can sometimes be hard and not everyone is 'genuine' even if they seem like they are at the time. Families are the most important thing in the world, more important than anything else in the world; you shouldn't cheat your families as they are all irreplaceable and precious nor should you underappreciate them, because you may not get another chance. I only wish everyone had a supportive family like me, it would help my BeatDepressionTogether target audience a great deal.

I wasn't sure why I was getting these very random thoughts as I didn't usually get them or think like this, perhaps it was because the fear was going as I was enjoying myself. I just couldn't understand them, then again it could be me finally beginning to understand the big wide world a little bit more. It wasn't clear.

We went scuba diving in Costa Calma and then we watched my Aunt and Uncle go in a submarine. It was a new journey and adventure for them, Marcus and I stayed with my Aunt Emma, the one married to my Uncle Tim, not the other Aunt Emma who is engaged to Uncle Mark, and it does get confusing having two Aunt Emma's! The submarine gave them as much enjoyment of the holiday as I was having. It was nice to see other people enjoying what I enjoyed too. The scuba diving was incredible, amazing and fantastic. That describes how I felt about the whole experience. I was really enjoying myself out there and being me for a change instead of being what felt like a big fear was nice.

As I had never been abroad or done any of these amazing activities before, I felt like I could carry on being out here. I was sure our time abroad would stay still but I couldn't believe that…

…The first week of the holiday was almost over! Why does time always fly past us? We never really appreciate time, rather get annoyed with the quickness of it. It didn't bother me though as I was having too much fun to worry about going back home and the holiday still kept getting better, we were there for two weeks, much better than just one week for sure. I didn't have a set bedtime on holiday, so I could stay up gone past nine o' clock and enjoy the resort and the great entertainment it had to offer and I sure did do just that. I was dancing and singing on stage, everyone was laughing and clapping and even joined in. I still had my great choir voice then, but I had to have a bit more of a modern voice rather than a hymn voice approach to the songs. I was so shocked and amazed how friendly the staffs were and I really was beginning to feel on cloud nine.

After singing on the big stage, I went and found my cousin. I wanted to go and continue to have fun so my cousin, I and both of our friends also called Matt, funnily enough, all decided to go and have fun on the resort; so we went around the hotel to see what we could do and we decided that me and my friend Matt would verse Marcus and his friend Matt in a game of hide and seek. We all agreed to play this game and we hid around the hotel, it was dark, so must have been evening time, but there was plenty of lighting around. The hotel was big so we all knew it was going to be a big adventure to find the others but none of us cared, it was all fun and games. Although the hotel was big, I was sure my hiding spot was the best! I suggested that we hid in a lift which took us up and down from the restaurant, it was very fun and we weren't found for hours until my cousin told my Aunt that we were missing. She obviously panicked and must have ran around the whole hotel looking for us both. I didn't have my mobile with me so she couldn't have called to see where I was. She must have been out of her mind and then she found us, it was very amusing that we were found after so much time. The way she found us was a massive shock however, we went down in the lift and then the doors opened and she was just stood right outside the lift with her hands on her hips. I was made jumped and screamed. It was a little embarrassing as I wasn't expecting to see her there, Marcus and Matt was stood behind her and was laughing. She must have been waiting for the lift to take her upstairs to search for us. My Aunt however did not look amused; she talked to us about possible fires occurring, safety and of course, staying together and not to muck around in lifts, instead use the stairs if we need to go upstairs. This all made sense about using the stairs. My pal Matt and I was just happy that we won as they hadn't found us! I remember Matt informing me he lived in High Wycombe, near London I believe. I didn't take his address to keep in contact, which was unfortunate because he was a great laugh. I haven't seen him since this holiday in 2005. Although we were lectured a little, that happening did make another great memory seeing my Aunt standing there waiting for us, I still remember that shockingly funny moment, I'm sure she does, but whether she chuckles about it in her own time is a different moment.

After the game of hide and seek, we all felt hyped so decided that we would explore some of the pools in the dark, so we went and had a great look at all of the several pools, each of them were quite large and had lights in, so looking at them in darkness was quite picturesque. After that, we went and had drinks, I continued feeling happy and enjoy both Matt's and Marcus' company, it was great fun and couldn't have come at a better time for me, in this moment in my life.

The holiday was never boring or dull and the island was one that I want to return to, sometime soon I hope. A holiday is the best thing to cheer you up, my fear seemed like it had gone away and I could make the most of what I had before it all had gone for good, the holiday must have been great because when I look back to that moment, I get a good feeling and a smile. It was so memorable. It would have been good if my mother and father had come along too, but mother had given birth to my little sister this year, so they couldn't fly with us, but nonetheless, it was great being in a different place, with different people.

After we had this fun and played hide and seek for hours, we decided to retire and went to get my Aunt and Uncle, Emma and Mark and asked them for the key card so I could get to bed. Marcus got his Mum, Emma and Dad, Tim, to take him back. Both Matt's also got their parents and asleep we were.

When we woke, Emma and Mark insisted we visit the beach early, it must have been around 6am. So us three headed down and I saw this amazing, vibrant sunrise. Never had I seen a sunrise anything like this. Mark really does love geography and learning about the climate, so his passion was to be admired here. Emma also loved seeing the sunrise and it really was worth doing. I mean, how often do you see a beautiful sunrise above a beach abroad? After this, we headed for breakfast, because it was all inclusive I felt the need to fill my plate, the food was lovely, had it not been, I probably wouldn't have filled my plate up. Then they told me, that I had been screaming all night, I thought they were joking.

"Screaming? Ha ha, are you two having a laugh here?" I remember asking in a puzzled tone.

"No, we thought that somebody had broken in, Uncle Mark got up to see what was wrong with you, and he thought maybe an intruder scared you." Aunt Emma replied.

Never has this odd occurrence got out of my head. Could this screaming this night be due to a bad nightmare I was having, or because of what had been happening back home? I just didn't know and even now I cannot answer this, it was very odd for sure.

That second week of the holiday, again seemed like it flew by and my friend Matt left that week days before I went. I said thank you to him for being a great friend and joining in with us lot. He pretty much said the same, he was happy he had met us guys, then he spoke to me about the school transition and he was experiencing the same problems as well, I felt like the talk with him was actually solving my problems before they happened. He seemed to understand my nerves and because he was at that age too, it really felt comforting. He was also worried but not fearing the move. Infact, I believe he was looking forward to meeting new people, something that happens so often in life, but how could I look forward to meeting new people, when I wanted the people I already had with me? Again, the difference in everybody showed up here, he may not have been worried, but I was. It did make me begin to think about the whole transition process again, I only wish he hadn't spoke to me about this.

His time to depart came and after waving him off, I went into the restaurant and looked at the camera I had, all of the great photos that I had taken of the holiday, even before the last day I wasn't dreading going back home, because the chat with Matt seemed like it must have helped and I knew I could beat my fear, or at least try to beat my fear over the transition even if it had come at the wrong time!

I just looked at the photographs on the camera and remember thinking; I wish I could live here. But deep down I knew I could live there and that would be an easy way out and easy doesn't often happen. I pictured myself living there though, that didn't stop, the weather was better, the people seemed friendlier and the whole atmosphere was great, but that's probably because holiday-makers want to enjoy their holiday, besides I couldn't live in the hotel forever. **A few of my holiday pictures are below:**

Here is the beach, Costa Calma. It was lovely here.

This was one of the swimming pools onsite, it overlooked the sea and was great to be in.

Here is what we woke up to every morning, clear skies and sunshine. The palm trees added to the admiration for this place.

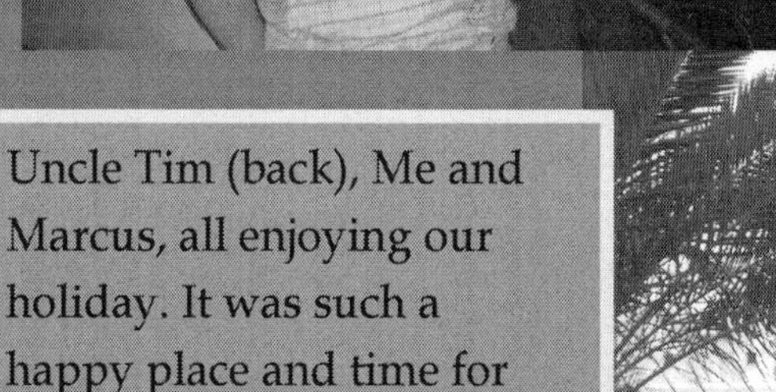

Marcus (right) and I smiling for the camera. A smile can hide a thousand fears, but this picture shows us both happy.

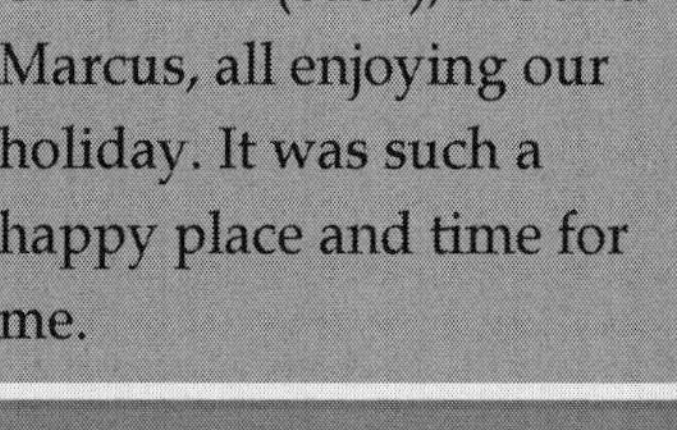

Uncle Tim (back), Me and Marcus, all enjoying our holiday. It was such a happy place and time for me.

And here we all are, minus Aunt Emma and Uncle Mark. Me, Aunt Emma, Marcus and Uncle Tim, enjoyed our holiday.

Aunt Emma and Uncle Mark on one of their adventures on the holiday, both looking young, care-free and full of enjoyment.

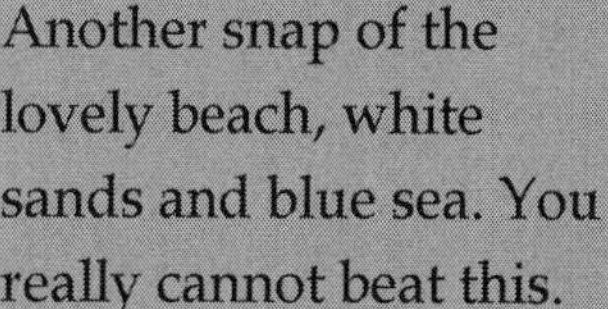

Another snap of the lovely beach, white sands and blue sea. You really cannot beat this.

The last picture I want to show you, here we all are on the beach, Mark, Emma, Me, Marcus and Emma. Uncle Tim took the picture.

My mind suddenly changed though. On the last day of the holiday I didn't want to go back home, it was hard to pack the suitcase. Even though I had that chat with Matt and felt I could beat this fear, I just didn't want to go home, so I went down to the beachfront and threw stones into the sea; I wanted to stay in the Island forever and ever and I did dream about this as mentioned, when you get something into your brain, it can be hard to get that thought or thoughts out, in this case it really was. Marcus must have seen me and followed me down to the beach, it was just down a few steps from the hotel. He told me that I had to finish of packing the suitcase and he would help me do it, if I wanted his help. He was sent in my time of need, again! It did feel like he was giving me support at the time, but perhaps he was asked to do it? Nonetheless, he cared for me, that close cousin relationship we had then, we still have now.

He was a very thoughtful kid, he had been raised well, but he liked me, did have a desire to get the latest toys, games or gadgets. I accepted his help to pack, after throwing a few more stones into the sea, I wished I was throwing the bad times in the sea that would have been better. We headed back to pack my suitcase and it was a case of taking our time to make sure we didn't leave too quickly, I didn't want this fortnight holiday to end in what seemed like days! I told him, I will be back here one day mate and I'll take you. It seemed that easy. I haven't been back since, I wouldn't go alone anyway, I just couldn't, but still, maybe one day I will revisit that lovely place. Everyone else was already packed and ready to go and we were still packing.

I was taking my time but it didn't make a difference, we were still leaving on that day, my suitcase didn't make the difference, neither would my actions! So after eventually packing, we left the room and went into the main reception area and carried our luggage down. It was like a walk of shame and an upsetting one, I remember looking around the hotel and seeing new people arriving and queuing up at the desk waiting to be booked in, the hardest part was seeing the lift that brought back all the good memories that I had whilst being on my first ever holiday abroad. The saddest part was that we were going. Another place I wanted to stay was going to be taken away from me, I was so happy here and my fear had gone for that time, well it hardly appeared, only once or twice.

Waiting for our taxi was a nightmare, the taxi driver forgot to pick us up and then it was a rushing time. So the time we could have spent on the beach had been taken. The taxi driver was called again and he made sure that he got us to the airport as quickly as he could and it seemed like this happy time was well and truly over, it certainly was not going to last forever anyway. Happy memories were made but time goes faster than ever before when you are having fun. I had met a new friend who seemed like me and Marcus had found a new friend too. I didn't enjoy the ride back to the airport and I certainly wasn't going to make out I was happy going back. I recall looking at my family and they looked happy and I heard them talking what they were going to do when they got back home, I wasn't interested in looking forward to being back in England; I wanted to stay abroad on the hot islands without worry.

The saddest part of leaving Fuerteventura, the dream holiday, was that I knew that the enjoyment I had from the holiday was going to be gone and lost forever, instead it would be memories, like everything in life becomes. I was a completely different person abroad, I felt I could do whatever I wanted to, walk on the beach at 6am in the morning and eat when I wanted and meet new people and sit in the pool. I could have a laugh and joke, I didn't have to get up and go to school or worry about school. It was like heaven! I felt young, wild and free.

The rushing wasn't an advantage though as our flight was delayed until 9pm, meaning we wouldn't arrive back into the UK until at least 2-3 am. I didn't want to go home and the delay was making things a lot harder for me. I already didn't want to go, so waiting around in the airport wasn't going to help. It was like someone from above was playing with my happiness and depriving me of enjoyment that I needed most as a child, even though I went on a lovely holiday. I was getting more negative thoughts about the future waiting in the airport and I started worrying about not being able to find new friends or get along with any of the new people in my new school, I was worrying again, fearing the worse brought out the worse in me, it still does now, but I couldn't understand why I was becoming like this. I always used to be happy, way before the 'big fall-out'.

I had my camera in my pocket, so I decided to get it out as my pictures of the holiday was on it. I decided to have a look through all of the pictures this time, rather than the few I had taken when I last looked in the restaurant, so I spent most of the time waiting for the aero plane to get into the airport looking at them. I was reminding myself of the good time I had. I hadn't really brought anything that had the memory of Fuerteventura for me, more for family members; I couldn't bring the beach and weather back with me and that was all I wanted, so I only had the photographs. The plane finally arrived around 9pm and we all got onto the plane, it was a scary thought returning back home, another day almost gone and another day closer to the transition.

We finally got back to England around 2-3 am, it was very dark and the airport was full of people, it didn't seem real. My family and I was walking through many areas of the airport and eventually I saw my parents. It was nice to see my parents again though; they waved and shouted over to get our attention. I ran over to them and I realized that I had missed them more than I thought. Just being back in England wasn't as a bad as I thought… at first, at least I saw them.

Questions started flying at me.

"How was it Matt, how was the holiday?" was the first question my mother asked.

"It was great Mum; I really enjoyed it and made some great memories, look at my skin colour, I have a great tan!" My skin usually stays white in this country, I envy those with olive skin, who can tan really easily, I wish my skin was like that, but a way to describe my skin is that it goes red, before brown and often, this takes time.

I enjoyed telling them about the holiday. It was so great to relive some of the memories and not being there, although I did miss the place and I remember thinking that I will elope to Fuerteventura and leave England at some point, it was madness really. I couldn't just up and leave and I certainly didn't do that even now I wouldn't go back there alone and certainly wouldn't live out there. We all got into the cars and drove back home in the darkness, it did feel emotional for some reason and I knew the holiday had finished, but at least I got the opportunity to go.

I was glad to get back home after all of the waiting at the airport I had to put up with. It was nice to see my sister and sit down and chat thoroughly to my parents about the holiday. I showed them the photographs of the holiday and some pictures of the holiday made them thrilled.

I could finally sleep in my own bed, in my own room! Being back at home was nice; I could see all of my possessions and the rest of my family who had missed me. I didn't realize how missed I was, I thought everyone would be happy for me to go on holiday and give them peace and quiet as I wasn't a quiet person when I was younger; I was very outspoken and always spoke my mind. My opinions mattered and everyone had to listen to them even if they didn't want too. My family always listened to me despite me not thinking they did. I was always close to them, which made it easier to talk things through with them, but not talking certain things through with them, did make life harder than ever before, my worries for instance.

I went around some of my family members and started telling them about the holiday as well as showing them my favourite photographs which were taken. I was reliving the holiday but in memories instead of presence. It is an odd feeling seeing some place that you have been too, when you're not actually in that place. It was again nice to tell the story of the holiday and the photo's helped bring it all back to life. My Grandparents were happy to see me, it showed that they had missed me as I hadn't been round in a while. They were glad that I had a nice time, but deep down I know they missed me, some of the things they were saying gave that away. I have always been close to them.

I couldn't believe how quick the holiday went, I didn't care how quick it was going when I was on the holiday as I was having so much fun, but being at home felt tiresome and boring and it is never a good combination. I didn't have anything to do or any friends to see as they hadn't even visited whilst I was away, I think I only told a few close friends I was going away, but they hadn't knocked, unless they remembered me telling them, which was probably the case. This was the longest time since I hadn't seen any of my close pals and it was hurting me inside, I didn't tell anyone at the time though, they must all have their own worries and be pre-occupied with their thoughts, without me causing a burden to them. So I just didn't bother any of them and sat at home alone.

Being at home still waiting for the transition from primary to secondary school was becoming really hard for me. I tried to occupy myself by playing my favourite computer game, The Sims 2. Life was easier playing that, I had control of My Sims' lives and I wanted to control mine as easily as theirs. The clock seemed to be going slower than ever before though. The holiday went quicker than this and the days I did nothing at home went slower and slower as each day passed. Only being a few weeks into the summer holidays, it felt as if I had lost contact with most of my class from primary school, my friends hadn't spoken to me and most of my old class were not going to the same secondary school as I was, so it was very hard. This thought kept recurring, it was bringing me down.

Having so much time on my hands without making contact with my friends or even having my friends to talk too, made it annoying but at the same time some could say I didn't put in the effort with them either. I could have possibly done more to see them but I felt like I was going to get rejected like I did in primary school, I was changing, I couldn't stand up for myself or see anything how I saw it before this argument. Perhaps I was still hurt. I felt my life had changed since getting back from the holiday.

The transition was coming around quicker than slower. I knew I had to start planning for the move but really couldn't, I felt weird and kept getting headaches and really wasn't approachable due to mood swings and a temper, that was very uncontrollable. To add to all of this, I was going through puberty, putting on 'puppy-fat' and becoming very self-conscious. I had never had a temper so this was something new and worrying. I was very mixed up over what was best, and I remember thinking everyone was out to get me, that is why they are sending me to another school I believed. Of course the move from primary school to secondary school is a move that everyone has to do so I shouldn't be any different; but it seemed like I was the only one that didn't want to go or change...

Chapter 6- Secondary School Begins

I hated having a temper, it wasn't a great thing. I kept feeling anxious and nervous, as well as being very self-conscious, I hated my image! The weight increase did really bother me for years, I must have weighed around 15 stone and it wasn't over-eating that caused this either, which I could understand. It appeared the worrying and going through puberty contributed towards the weight change. My mum couldn't believe how worried I was, I think she knew already, but didn't know I was worried as much as I was, she thought that I was pretending so I could stay at home for a few extra days but it wasn't that at all. I just didn't want to move and I wasn't going to go to this new school being the happiest person there but at the same time I wasn't going to let my fear win, I couldn't let the fear control me, even though it was. I kept telling myself, as I have told you that I would beat this fear and trying is what I did.

To try and cheer me up and get my mind thinking positively, mum decided that we would go shopping for school essentials. I got all new stuff for my new school, I got the uniform, a new bag, a pencil case and other stationery items, it was odd seeing the new uniform, infact surreal as I still had my other one hanging up in place, waiting for me to wear it. Of course, I wouldn't be wearing it again! The new uniform didn't seem as nice as the old one, it looked dreadful, and it didn't have the St. Mary's school logo on, but instead the Longsands' logo. I tried it on and immediately I didn't like it, it wasn't me and it was a vile dark navy colour. I don't think there was anything wrong specifically with it, rather the fact that it wasn't what I was used to wearing or wanted to wear. I hung it up at the back of my wardrobe and didn't bother to think about the uniform again. After shopping, I thanked mum for taking me, but it felt as if this was one of the last steps we had to take before my new chapter, the new school.

I decided that I wasn't going to spend the last few weeks of the holidays moping around, so I visited Josie and apologized for the last visit. I admitted that I was scared and feared the change as it wasn't what I wanted. I then expressed my thoughts and told her that I wanted my old friendship group back and I didn't want to make new friends who didn't know me, they didn't see me grow up from a toddler to a mature child, but instead they would only know me in teenage years. She looked at me and then brought out a photo book, it was her back at school as a child, something I had never seen before!

She showed me the photographs of her as a young girl and told me that everyone has fears, it doesn't matter who you are or what you do, you have got to confront your fears or they will win and destroy you. She confronted her fears and although schools were different when she was younger, she was nervous and scared of her first day at school. She said it wasn't going to be easy but I was making it harder by judging it before I started. I could see her point, it hadn't happened yet and I was already nervously worrying and had been worrying since the middle of Year Six at least.

Josie had kept some of her school memories; she kept a piece of her friend with her all the time. She had a great smile and it was good to see her happy memories as well as hear about them, every person has a story behind them. It looked like she was going to have a tearful joy, so I knew she was telling the truth about her emotions and how she felt. Deep down I knew she wouldn't lie to me anyway, I trusted her advice about school and remember thanking her and then she took me back home, she walked me back with her dog, who she adored passionately. Within a few days, I had a card from her wishing me good luck in my new school. It felt great to have this support.

Once home from Josie's, I got my new uniform out from the back of my wardrobe along with my old uniform. I put them both on my wardrobe door and started looking at them both, it was as if I was comparing. I got my old uniform and decided that I couldn't wear it again, there was no need in wearing it again after-all, so I decided that I would throw it out in the trash; I sorted through all of my old school uniforms and got rid of them, I don't think they went into the trash though, I believe they went to Marcus. I then sorted out all of my old school work, some I kept and some I threw out. Some had loving sentiment to me, even if others didn't see it and just saw a few doodled lines. The time I threw most of my old school things out felt difficult, as it was like I was cutting the apron string from wanting to be attached to St. Mary's. After throwing the lot out, I felt like I was setting myself free from the school, I could at least continue feeling the memories as they hopefully will stay with me for life.

I then tried the new uniform on again, I brushed my hair and put on the new shoes I had really got, after the shopping trip with my mother. Once I felt ready, I turned and looked into the mirror with shock. The uniform actually looked different than what it did before; it didn't look vile or a horrible navy colour. Instead I looked smart wearing it but didn't think that it was me. I hung it up again. I really wanted to let go and welcome the ever-close transition, but I couldn't. Little did I know, that I was to become depressed and change forever would happen, looking at this now, I should have known really, but little was known about depression at that age, we were never taught about it at school, perhaps it could be gently added to the curriculum now? Mental Health has such an importance in this society and needs further understanding, development and students need guiding. Perhaps bullying against sufferers would stop then?

The final week of the school holidays was here and I was tired, so tired infact that I was suffering badly from exhaustion and felt ill, worse than ever before. I think that all of the worrying I had been doing had finally caught up with me; I was very comprehensive of hate towards this school and wouldn't accept that in one more week, I would be joining that school for good. It didn't seem right and it didn't seem fair! After the two day visit to that school, it felt big and didn't feel like somewhere I could imagine myself, it wasn't a small, dainty school what I was used to, it felt like a huge maze. I hadn't even started, but yet the feel of hate had already began, it is an unimaginable feeling, unless you've been through the same worry as I.

Feeling worried and having nothing to do really did impact me, I wanted everything back how it was, and nothing ever did go back the same. I feel that this is a huge problem for many young people who go through this change, they fear the transition as it happens so quick, yet prolonged if that makes sense? They should be guided through the process, both schools, the existing and future one should work together, in order to make sure each student feels happy and any negative thoughts worked on, I do honestly feel that each child should have this support, otherwise they will be like I was, full of worry and suffering from what I now see as depression.

The school holidays were up. The first day of my new life began and my last free day went quick and before I knew it, I was getting ready for secondary school, days, weeks and months' worth of worrying over.... Mum came into my room with breakfast and woke me. I didn't want to get out of bed as I feared the worse. My alarm then went off and I accidentally smashed my cup of coffee on my bedside table, the morning was already bad and that's before I even got out of bed. When I did get out of my bed, I got washed and dressed for school. That was when I realized my trousers were too tight around the waist and they didn't fit. I had to breathe in for them to fit. It was awful and I was rushing furthermore to get to school early. I knew over that summer break I did put on a lot of weight, my family said it was 'puppy-fat', but no matter what is was called, it was still something I didn't want and certainly didn't feel comfortable or happy about having. I guess the worrying added this fat to me? Or maybe it was 'puppy-fat'?

The first day at secondary school was an experience that I will not forget. Having the worry of worry, I felt little happiness. I had feared this moment since the middle of Year 6 and all through the holidays, as I have shared with you. I was already nervous, my nerves were with me on that day and they did not help me, not in the slightest, those who say nerves do not affect a person's wellbeing are completely wrong, nerves can stop so many things from happening in life, again at the time I didn't realize I would be diagnosed with depression and anxiety later on in my teenage years. I didn't want to meet any new people or leave old chums behind. I packed my school bag and walked to school, I feared this moment, walking into what seemed to be a new life for me and leaving my old one behind, and then I got into the playground and all I could see was queues of people, there must have been hundreds of students my age waiting, most whom I did not know. I was panicking even more, in my old school there wasn't even a hundred students! I just remember getting this feeling that told me to walk out of the playground and straight back home, but I was determined not too as Josie said confront your fears and I wanted to at least try anyway, so I continued walking down this long road in the playground and eventually I saw my tutor group and a tutor. There was a sign saying JB above my tutor, so I assumed that was my group. The group seemed ok at first, all of them had different looks and some did know each other, some didn't. Most of them looked nervous like I did, but the thing that got me the most was that it appeared that most of them knew each other or had been introduced to each other, some even went to the same primary school as the others; I didn't know any of them! Most of the group, were all laughing and joking together, they turned around and looked at me when I queued up, I just put on a smile and looked around them, I was very inquisitive and tried to figure out who I would try and make friendships with, it wasn't easy, because not knowing any of them made this a challenge.

The new school surroundings were much bigger than I ever had expected, I did have the two day visit as mentioned, but for some reason the school and its surroundings felt massive. And the year groups were much larger than ever before. After the long anxious wait in a queue of at least 20 other pupils, we were finally introduced to our new tutor. She was very different to my previous tutor as she appeared stricter, more belligerent and didn't have the knowledge of me as my other teacher did. Mrs. King was calmer than what she was and didn't have this atmospheric vibe from her. One way to describe the feeling was like going into a new family and not having a clue who they are or why you were there for. I could tell that this was a door to something new and that everything was about to change. She seemed the sort of person that would judge a book by its cover, she didn't even smile and just told us to follow her. She appeared lost in a world of her own and this wasn't good. I liked to know my teachers and have knowledge of them, she didn't seem attached to emotion, but felt cold. I won't mention her name, because we didn't really see eye to eye, but she had blonde hair, which must have been dyed because she was in her late 40's, she was around 5 foot 4 I would say and had designer clothes, she also appeared to care about her look because she was focused on how she walked and talked. Very self-conscious I would say now, having experienced this.

Whilst we were following her, in an orderly queue, I believe a surname alphabetical queue, I remember getting these strange feelings in my stomach, I just thought be strong, you can do this, so I just continued smiling at everyone I met, I didn't want everyone else thinking that I couldn't cope with the transition at the time. If they could, I could. Why was I having all of these problems? Whilst they were laughing and joking around, it wasn't fair and something didn't seem quite right. I have always been the sort of person that puts feelings in a box and put on a strong face that would get me through the day. I didn't like showing my emotions as I felt embarrassed and anxious when I did. Also the worry of people judging me or not understanding did have impact on me. Now, I couldn't care what people think about me, I have been mistreated, misjudged and misunderstood many of times that the fear of being judged, just passed. It is a worry that many have though and it can easily destroy self-confidence, especially in those with mental health and depression. Little is understood by others and one judgment can affect that person for a long time, I remember somebody saying something cruel to me about my weight, I took this personally, I took a lot personally and to heart, so I can relate to others, especially those who have ill-feeling towards judgment or have been judged themselves. Nobody has the right to judge others and nor should they! I couldn't tell whether the other people in the tutor group were feeling the same as I was or whether they had bigger ego's than I did, which gave them a confidence boost to keep on laughing and putting on braveness. After a short walk, we, the class, all sat down on our designated tables and we all had followed a seating plan. Once seated, the tutor then asked us all to make an introduction to both her and the other class pupils. She started on my table first, her deep green eyes looked all around the classroom and at us and then her eyes focused on me. I was one of the first to introduce myself, I remember saying nervously:

"Hi, I am Matthew and I enjoy swimming."

My throat went dry after introducing myself, but luckily as we all had to introduce ourselves, nobody judged me and at least I wasn't the only one that had to address the awkwardness of not being known in that class. That was the only thing that appeared in my mind. I didn't want to have a friendly chat with people I didn't know on the first day, it didn't feel right to me. In-fact it took me a while to introduce myself to many people and the reason behind this was still unknown. I was quiet, bitter and resented the move deeply. Something I had never been before was quiet and certainly not bitter.

I had the same feeling as I did in primary school, I wouldn't knock my pride and my pride was never going to allow me to just introduce myself to anybody. I call it my pride, because I am a certain way, I am different to your 'normal' person, which may I assure you, does not exist. It took ages for me to properly introduce myself to some of this class but by the time it took me to introduce myself to everyone, they all had their own little group of mates or had other things to do and I didn't. I remembered seeing someone who just looked like they had the same feeling as I did and I expressed what I thought:

"I hate this; I just want to go back to my normal life."

Their reply was "This is MY normal life; this school will be your second home but it won't be mine!"

It seemed as if they didn't want to be there anymore than I did, but what could I say back to that?

The conversation ended promptly after they cruelly said that to me. I didn't want to listen to them judge on what would and what wouldn't become of my future, I didn't want to attend the school and I certainly didn't want it to feel like a second home. I had to find something to do to stop me worrying. I just sat down and finished listening to everyone else introducing themselves. Little did I know that the conversation I had earlier was actually another sign, the person I thought was insulting me was actually telling me that I had to change and get used to the new school as it would be a second home, I would be in the school 5 days a week, every week and learning different subjects with different tutors and students. I just didn't see many signs at this young age, they were all present and had a funny way of presenting their faces, but never at the time, did I really understand them. Ignorance is bliss as they say.

After all of the introductions and a break, we all got our timetables that showed us what subjects we would be learning; at that stage it was fine. Our class would stay together as a whole and be learning the subjects together. This was fair enough and was a good way to make a bond with everyone in my tutor group, staying together was good whilst it lasted. In a way I was glad that the school kept our tutor group together for the first year otherwise I probably wouldn't have bonded with anyone… but from then after, it wasn't all fun and games.

Since allowing fear to take control of some things I did and who I was as a person, my personality had faded slightly and I wasn't a strong-willed person anymore. I wasn't able to think positive thoughts, I always spoke negative. I couldn't control what I was saying and that certainly did have problems for me… it was very unlike me to do this, I loved seeing the good in everything and everyone, it just wasn't explainable.

We couldn't all sit around doing nothing and our tutor took us to our first lesson. The first lesson of the day was Mathematics. We had yet another introduction with our Mathematics teacher and she told us that we would be studying many areas such as decimals and algebra. I couldn't do either of them, I still am no genius at these subjects, and it does make me feel resentment towards the subject, because I cannot understand the subject. I didn't understand nor could cope with them at that time either. The topic was Algebra after all and that was a tricky topic for me and when nobody explains a new thing to you, how you need to be taught it, how are you supposed to work it out?? Everyone has different learning styles after all and mine wasn't reading a worksheet to work the answers out. I just sat in the classroom in silence and looked around to see some other people looking into space like I was. The first lesson was hell, in my opinion, as I had never done algebra before, it may have come up once or twice in primary school, but we didn't use it on a daily basis. Why would we have, it is not as if you need algebra in the future, unless you want to be a mathematician. I hated the first subject and I certainly wasn't looking forward to the other lessons after that, it seemed as if my gut instinct was right and perhaps I should have stayed in bed…but this wouldn't have achieved anything. I would have continued getting the anxious, fearful worries and it would have taken my mind into overdrive again, I hated these feelings.

Fortunately, there was break after Mathematics. I was so glad that the lesson from hell was over; I hated Maths as I couldn't do it in primary school either. I was always better at Literacy and enjoyed that as the subject is a world of history such as Shakespeare and other amazing plays and poems, as well as non-fiction and fiction stories. After Maths, I just followed the crowds down into a large daunting hall, it did have a name but I have forgotten it; I decided that I would sit down waiting alone for the next subject. I was waiting for someone who I could talk to and have a laugh with, to come up to me and make a conversation. I liked to know I could trust people before I allow my friendship with them, I do find it hard to have people in my life that I cannot trust, it has become an important factor for me now, I have to trust them and I want to know they trust me, otherwise I wouldn't have them in my life. I was sat there, on one of the stairs and kept looking around at everyone else talking, the hall seemed bigger than what it actually was and I felt frightened to be alone in the hall. The loudness of everyone's conversation was overwhelming, there were many conversations all going on at once, something I've never had to endure before or since, luckily.

As I was scared, frightened and worried, I was all alone. I was probably showing my loneliness by not being able to stay still, I had a rapid movement in my leg, where I would just move it up and down whilst sitting. I turned my head around the hall and saw a few members of my class. It wasn't until I stood up and went to exit the hall, that two people came over to me and invited me to join in their group. I thought that things were on the up for me and that I was getting noticed despite my confidence dropping, I did my utmost to hide this, some people never knew I wasn't confident, even when I told them years after, they couldn't believe it. I did put on a 'fake-confidence' so others didn't think any different of me. I was glad that they were in that hall with me, despite the dreariness of the hall; I wouldn't have hung around with them otherwise.

The group at the time had 3 members: Ben, Kathryn and Melyssa. Mitchell joined later as he started the school later than what we did. Tina also joined the group, soon after. The group was very kind and friendly towards me and I was over-whelmed by the kindness that they were showing to me. I hadn't been shown any kindness since joining the secondary school and I missed it, it had only been a day though. They seemed like good genuine people, not the sort of people that turn around and gossip about you, or stab you in the back the minute you turn around. It felt a great sense of relief to finally be in a group of people who I could talk to. Anything is better than being alone anyway. Loneliness often is a part of life, I experienced it and do still experience it sometimes, I feel and felt as if nobody understood me and that I was alone during my tough times, loneliness can make a person do odd things and act out of character, it can often lead to a great deal of emotion, which is why the importance of close friends and families is vital.

Kindness is one of the traits many do not have! I have met many people over my 20 years existence and I wish I could tell you that they were all kind and supportive to me, but sadly I cannot. Being kind costs nothing and is positive. Being rude, unkind and unsupportive, constantly mocking or trying to degrade someone with a different, is negative for sure. Kathryn, Melyssa and Ben were real people and seemed like they were going to be good friends for me. After the introduction to them, we all spoke and told each other some stuff about our pasts, our hobbies and interests. I was finally enjoying being with new people and thought the change wasn't all that bad, especially now I knew I have these friends around. I told them how I hated being in the big hall and then we all went outside, it was a lovely summers day outside, and we continued talking to each other about our lives and what we was looking forward to doing in the school. They all had loads that they were looking forward too: the lessons, meeting new people, getting new skills and going on trips. Then they asked me what I was most looking forward too.

I recall that after they had asked me, I just went blank, speechless infact- I wasn't looking forward to doing anything in the school. Thinking fast and quick on my feet, I just improvised what I was looking forward to doing and said I was looking forward to meeting new people and making new friendships. I wasn't looking forward to meeting people and building friendships though, but deep down that was the only thing I wanted to do, it just took me longer to realize this than most. I didn't tell them about my past problems with different friends or how I felt I couldn't trust many people a lot. That was private and I attempted to keep it that way, I didn't want them to know anyway. I wasn't going to judge their loyalty and support the same way as I judged attending that school, they did seem great people after all and I didn't want to ruin that by stressing them out with my worries and concerns and with what had happened that past year.

Building friendship was all that I really wanted to do in that school. The curriculum didn't motivate me at all, the subjects were all dreary and tiring, the subjects, well the topics anyway, were all new to me and I just didn't understand or want to learn them. You cannot teach someone something that they are not willing to learn and I wasn't going to learn new things that seemed irrelevant to me. I just couldn't get to grips with why the curriculum was changed so much, especially after just coming up from primary school. It was like another language and it made it more complicated for me to be there. I was already getting annoyed with not knowing who I could trust, I just met these new friends and it wasn't as if I could tell them my whole life-story, I didn't feel comfortable with the thought of doing that. Friendships were easier to understand than the new topics. Everything takes time and understanding the subject topics which they expected you to know immediately took me ages. Me verses Algebra, algebra wins! I certainly couldn't change the way I saw things, no matter how hard I tried or wanted too. This is where good understanding comes into things, if a teacher sees you are struggling, they should help and motivate that learner to learn, I didn't have much support, although I do sincerely thank Miss. Burt, she was my first language teacher and really did help me, she was a lovely, thin tall lady with much beauty. She taught me French and I thank her for that. Because I had never learned language, just like I hadn't learnt much about algebra previously. Again this is where learning styles arise again, every student is different, yet not every student is supported correctly.

Apparently change happens for a reason- some good and some bad. I couldn't see any change was good, everything I saw was bad. I thought everything I could see or touch would fall apart and when I returned home that day, my view hadn't changed, despite meeting these new people. The walk home from school was a longer route to the one I used to take, which was just around the corner. I had the pros and cons of that school written down in my notebook, the only positive was that I had met these new friends and as for the negatives, the list was longer than my Christmas list.

I got into my house and I was adamant that I would not change or change my views. I didn't want to go back to that school, I didn't enjoy that school much and I wanted my old school and friends back; I did love that school, the teachers were great and understood the children within. We had many enjoyable trips in primary school like going to the Isle of Wight and going to the local museums. I know the trips may only be memories and happy times but I wanted them back, I needed them; I saw positivity in that school! I couldn't accept that I couldn't have these times back, I needed them back at the time, and I wanted to relive that chapter in my life. Even though I thought I had accepted this, it would all come falling back in my face and acceptance certainly didn't happen.

People say everything will be ok in the morning or sleep on it and it'll be okay, but it wasn't. Forgetting my old school wasn't as easy as that though as I lived right near my old school and seeing it from the window every day, brought back all the memories. The memories kept that school alive for me and my brain was back in the past, it was looking backwards and not going forward to the future and present. What hurt even more was the sense of enjoyment I had out of that school. It took time to build that up from nursery to year six and it was cruelly taken from me.

Everyone was friendly, everyone knew everyone and the atmosphere was great. Now I was in a larger school that I didn't want to be in, I was with all different age groups, I didn't really like my tutor, I had a few people who I had acquainted but that was it and a curriculum that I didn't understand- what more could I want? I kept asking and asking for more help with the subjects I struggled with, but nothing was ever done for me. I was just kept in detention or given extra homework to see me through because of my 'disruptive behaviour'. It didn't help one bit though and I struggled to even attempt the work, let alone complete it. Eventually, I did get some support from Mrs. Young and the homework club that the school offered. Mrs. Young was a great Teaching Assistant, she supported me with Maths and had a great understanding, she was a short lady, with light blonde hair and a lovely personality, and she had much understanding of why I hated Maths and tried to encourage me to want to learn. The homework club staff, also teaching assistants, did their best to support me when I needed it. But Mrs. Young wasn't in all of my lessons, which was unfortunate, because if she was, I think things could have been very different.

I felt I was failing and I wasn't achieving. Again this feeling was another 'awful' one I was having. Achieving meant a lot to me, I had to achieve to get ahead in my life and to become the lawyer I always dreamt of becoming. I wasn't as able as some of those that were high-achievers in my class, but I wasn't allowing possible failure to deter me from achieving and I didn't bother doing any of the work until I had the proper help and support I needed, I was then informed of the club I just mentioned to you, - homework club!

The homework club was a great idea for people that struggled with certain subject areas or was struggling with their work load. I appreciated it and it was different to the class room. I wanted that to be the way I learnt, at my own speed and my own pace. Not how some teacher bosses you around to work. It wasn't fair to work at a high pace and not understand the work set out in front of you. I wasn't going to let a teacher boss me around and I wasn't happy with getting the extra homework! If you do not understand something in the workplace, you are offered support or further training, so why couldn't the teachers bother to offer further support? Perhaps their own attitudes got in the way? We aren't all clever-clogs at everything.

My thoughts and worry of failure, continued like this for many years after, even now I hate the worry of failing, but failing apparently helps us learn along life's pathways, so perhaps it isn't as bad as I once thought it to be? I still never had a nice thing to say about anything or anyone; I was very worried about everything and even worried about life which brought a tragedy to my health. I wasn't seeing things as clearly as what I used too and I thought that everyone in the school was judging me on everything I did and that people were talking behind my back and that feeling was one of the most damaging feelings that I ever experienced. You cannot turn your thoughts on with a flick of a switch, it is not that easy and I do wish it was, but the simple answer is it isn't.

That feeling never left me and it didn't matter if I tried to hide that feeling because my insecurities still showed inside and out; I lied awake with worry and thought that my thoughts and feeling were true, more sleepless nights for me to endure. Somehow they managed to seem real and I felt as if this was actually happening to me. It was very strange to see my fears come alive in my brain, but I wasn't thinking right that was the only possible solution. I couldn't change what I thought of people and it certainly was hard to pretend that I didn't think they were talking about me. It was like a knife getting stabbed into my back and being drawn all down my body, every time they asked me something, I would be short with a reply. A little reply, hardly any words from me was better. I had no evidence that my new group of friends was talking about me, I just sensed it. It hurt and contributed to my fall even further. The capability of being able to cope and trust had gone, I couldn't trust them and wasn't going too either. I didn't feel as if I was coping very well either, I was turning into a scared, quiet person, a person that hardly spoke to anyone and was becoming rather anxious and scared of everything!

After another sleepless night, lying awake worrying about what the second day would bring to me, I had to get up and dressed as it was school again! I was so tired and sleepy, that I dreaded going back even more. I entered the long road to school again and still had that terrible gut feeling. I was even awake before my alarm clock, as I did not sleep! When I got into school, the group came up and started talking to me so I spoke back; even if I wasn't a big speaker now, I always tried to put an effort in with them, my worries can't have been right, could they? It was quite an enjoyment talking to people and realizing that some did really have the same worries that I did as it made it easier to talk about it. So I told them that I was worrying about people and that they may be talking about me behind my back. One of them had the same problem and it felt better talking it over with them. They told me that they had a friend that betrayed them and then they got dumped for another person. It felt terrible listening to them as my problem seemed completely different to theirs; I wasn't dumped just ignored by my old best friend. The discussion eased my nerves but didn't take the weight away from my shoulders though. I was now worried that I had said too much, I only knew them a few days! I just was returning to a chapter in my past which I couldn't fix, I wanted to see it differently and for it to have a different turnout and my best friend accepting he was partly to blame, but it was too late.

They kept going on at each other, random conversation, like they had known each other for years before these few days, and the banter between them was good. However the pressure from them to talk to them and admit all of my worries was set upon me. The 4 of them, now including Mitchell, wanted to know more about me and I felt as if I was able to tell them some more about myself, more than before anyway. They wanted to know and I thought at the time it was the right thing to do, they told me that they were worried too and I believed them. I guess I heard what I wanted too, because they could have been acting on the long conversation we had earlier. All of the peer pressure and the change had confused everything I thought I knew and was.

I told them again about how much I hated the new school I was in and how I wanted to die rather than keep attending that school, my form tutor was standing right behind me, as I remember or was told, because she was ever so quick to appear, when I told them though and she didn't think that was funny, she seemed to think that I was having a joke when in actual fact I was being honest and serious!

Things got much worse before they got better though, I was very paranoid and that feels somewhat like an understatement. I was going further downhill and I could continue seeing bad things happening before they did and this wasn't normal. I kept going through the pain of being able to see things before they happened, despite nothing ever happening at the time I saw the fears. The worst part of that was actually going through the pain, I saw things that were not real like my old friends coming back together and my best friend was with them coming to save me, it was like I was living in two places at once experiencing these mind flashes. I couldn't understand why this was happening to me; I didn't have 2 brains or 2 lives after all and couldn't see into the future. It was like I was living in somebody else's life at the same time as my actual life and the feeling of paranoia and hatred got bigger and bigger as each day progressed. I hated all the hate for my new school because it took the old school away and it took the old happy me too. I wasn't able to hate as much as I could when I was at that school. Incase you are unsure what Paranoia is, it is where a person/sufferer has suspicious thoughts about a certain thing happening, like I did over the whole school transition and secondary school. Everyone who experiences paranoia is likely to be different and experience different thoughts or beliefs. For instance, yours may not be the same as mine, and vice-versa.

I never enjoyed the normal things that used to make me happy like seeing my mates or going out on bike-rides or even swimming, which did make me happy. My Grandfather taught me how to swim, he used to take me once a week to a local swimming pool, mother occasionally came, but I was enjoying bonding with my Grandfather, he's always been around for me. Something was taking my happiness away and I didn't know why this was. Those memories were taken from me as soon as I started to worry badly. The worrying wasn't needed or wanted, but it kept growing and eventually took over me. I couldn't do anything that the old Matthew enjoyed and even acted differently to him. I couldn't live how the old me lived, I just was completely different and I was like another person. Another person inside my body, living the life I hated. Everything about my life wasn't right or fair. I was very nervous about going out and I even didn't like my own body due to the weight gain and the acne I was getting, this wasn't wanted either! This fear was disabling everything that I was and it succeeded.

As the old me had gone, I wasn't able to talk to my old friends who I happened to see when returning from my new school, they all were together and there was me, alone without them. I never went out with friends after school, no communication with the outside world when home; instead I was in the house all the time, shut up in my room. Things never changed for me, it got worse and worse. All of my once loved hobbies all had gone, my kindness taken, I was no longer a determined, confident child, I was a worried, shy child which meant I couldn't enjoy what I used to and it wasn't me…nobody deserves to go through this.

I hated feeling like the outsider of my own life, why was I not getting the thrill and enjoyment out of everything I once did? I couldn't understand any of it, I knew for sure that I didn't want to attend the secondary school but was unsure on why everything was changing for me, surely it isn't what every teenager goes through, as the others all seemed happy even if they did have minor worries, I wanted their minor worries not my big worries.

I concluded that secondary school wasn't my biggest need; my biggest need was to be like everyone else, happy. Happiness is free, so why couldn't I be happy? I wasn't this nasty outspoken person that I felt I would be seen as; I never enjoyed doubting people or moaning about everything, it certainly wasn't a hobby. I wasn't a vile thinking person and that must have been what my tutor group thought about me, that I was twisted and jealous. They probably thought that I liked moaning and doing that as a hobby even though it wasn't one. I wasn't able to properly correct them as I couldn't even convince myself that I wasn't this belligerent, pessimistic, person who disliked everything in his sites. Deep down, nobody else's opinions should have mattered, I only wish I could turn off my emotion then like I could now, but I tell you, it isn't as easy as you think it sounds. I hated thinking that they thought this of me and again this was another thing I saw in my head and wasn't sure whether it was actually happening or not. I also hated not going out to my new group of friend's meet-ups, getting out having fun and enjoying the teenage years. It wasn't for me at the time; I couldn't enjoy school so I wouldn't enjoy seeing students from the school.

It was frustrating having to make up stupid excuses for not attending any of their events. I had an excuse every time they wanted me to meet up with them. I didn't want to meet up with them and found it hard to pretend I was even wanting too. I would have probably been happier if I went out with them and enjoyed their company but they wasn't my old friends, I still wasn't sure if I could trust everything they said and did, after all I had only known them for a few days.

Chapter 7, Secondary School Continues

It was yet another day at secondary school, by now it was about two years into secondary school, I remember this day well as it brought me more pain and misery than I had ever expected a school to do. I remember being asked to visit reception by my form tutor, I didn't know why I was the only person going, but I had to go. It was embarrassing getting called up to go to reception so I left the classroom quietly. I couldn't understand why it was only me going to reception and nobody else. I hadn't done anything wrong and thought it was very odd. This thought kept recurring in my head. When I got to the end of the corridor, I saw this tall lady, taller than me and I was around 5 foot 7 at the time, with glasses on, kind of half-moon shaped glasses, so I carried on and I walked to reception and it appeared she was waiting for me. I had no idea who she was and why she was waiting for me.

She introduced herself to me.

"Hello Matthew, I am Adrienne."

I remember wondering who are you? I had never seen this lady. I had never met her before and couldn't recall her being a friend or family friend. She seemed a very peculiar person at the time.

"Hello, you wanted to see me?" I responded in concern.

After going through my head if I had seen her before and realizing I hadn't. I was taken into a room with her, just off of the school's reception. It turned out that the school had contacted her as they were concerned about my health and well-being. The school apparently noticed me and wanted me to get the help I needed. I wondered what she was implying. I was only growing up into an adolescent after all. I knew that my family had noticed change in me but didn't think the school did as they never told me or helped with the problem they must have told her they thought I had. I sat in silence after she told me that, I thought and thought. Nothing was wrong with me, I wasn't the same person I was when I left primary school, no, but I was fine. I kept telling myself this, despite the fears I had.

She wanted to ask me a few questions; I didn't want her to question me. She asked me to give her an account on what was bothering me and what I was still struggling to adapt with in school and life in general. I couldn't accept the betrayal of the school, I didn't sign up for the school counseling session, I was getting enough counseling at home, I was getting worried as it wouldn't be long for somebody to see me with her and then my news will be revealed. Despite my worries though, I gave her the truth- I didn't want to be in that school, I wanted my old life back. I didn't enjoy the curriculum, I didn't socialize with anyone new and I wasn't getting out meeting the big world.

I think she listened to me and she seemed very interested in what I said. She then started taking some notes and continued asking me questions. She asked me if I was having any problems at home, I told her I wasn't. She then asked why I wasn't enjoying my life; I remember telling her that I wasn't enjoying it because I didn't want to be in it.

After our brief induction and meeting, she left and I didn't see her for a few more weeks after that meeting. It was very odd why I was the only one to have this meeting and I kept thinking, why me? I knew my positivity wasn't up but was it that bad and noticeable? It appeared that the counseling she was giving me was because of the way I was in the school and my outlook on school life and life in general. I jumped to the conclusion that everyone in the school must have thought I was really vulnerable and an easy target. I didn't want the news of the counseling to get out so I made sure I didn't say much.

I went back into my classroom and sat down in my chair. The tutor looked at me and she smiled. She must have known why I had the meeting; I wasn't sure whether she was being smarmy or nice, I just couldn't tell. The people on my table asked where I went; they obviously wanted some inside gossip but they wasn't getting any from me, so I just said I had to speak to someone on reception. That was left at that.

After not telling them about what had happened, I went into deep thoughts. I didn't really understand the chat I had with this woman, I had never met her before and I felt like I was yet again talking to another stranger, I couldn't grasp why only I had to go and see this woman from the whole class , I was losing my confidence, self-esteem and faith in life. I had already lost my positivity at this stage. This was a woman, a woman who didn't know me, she didn't know me when I was happy and she certainly wouldn't have any memories of me. Sometimes though, it can be easier talking to a professional, a person that you don't know as they will not judge you or your problems/concerns. I wasn't too sure if anyone else had been told about my worries over the school or not but I was hoping they hadn't. I didn't want the whole school knowing my problems, I didn't want them all to be laughing and joking at my expense over my fears. If they could cope with it, then why couldn't I? I kept going over this question over and over again for many months, which led to years and it didn't make sense no matter how hard I tried to understand it. Now, I realize it is because humans' minds are different, different reactions to everything, I didn't have the same as they did, not really.

My tutor obviously knew and the head of year must have known why I was sent to see her. My fears were already shared with this woman and the problem was with me all of the time, so it wouldn't have been that hard to see the problem. I didn't want anyone to see that I was weak, vulnerable or cold. I was lost inside of myself and it wasn't the same. I was quite the opposite too before this happened, I loved life and was happy. Now, I was vulnerable and felt weak because this change happened. I feared the news of my worrying health getting out to everyone and I was sure that it wouldn't be long until the school staff started gossiping and my news may be discussed around the staffroom, who knows who could hear? Fortunately, there was a confidence in place between Adrienne and I, she would never have discussed it with just any member of staff, and there was a line the information went down, kind of like a chain of command. I just wanted to keep the problem to myself; a problem of fear was mine and nobody else's. I couldn't trust this tutor and she certainly didn't trust me, she was so offensive towards me, she really didn't care for me and she certainly had no compassion for secrets, she told everyone else's business to anyone, which is why the fear of everyone knowing got bigger in time. She liked embarrassing people, I only hoped she got embarrassed. Perhaps she did as a child, they say a bully is bullied at home? But I sometimes would change that opinion to a bully is what becomes of upbringing. Then again, everybody has different opinions on what causes bullying.

I couldn't just drop this fear and it got worse, my fears were like a vicious circle, one after enough, also with regrets added and little understanding of what was happening to me, I just couldn't cope with not being able to see what was happening as clearly as I needed too. After that meeting with the woman I didn't know, I went back to my home and told my parents, in particular my mother I recall, about what had happened on that day, they must have known about that meeting as they didn't seem shocked or bewildered. I remember them just looking at me and my mother had a letter behind her back. I was betrayed and the school was informed of my business because of my mother's concerns about me. Why was everyone questioning my health?

I couldn't understand why everyone thought that I was turning into a bad example or had bad health. My health was fine and I wasn't going to let other people tell me different. I just couldn't understand why it felt as if they all had it in for me. I wasn't unstable and I wasn't about to let my worries take over everything; at least I thought I could cope with them, but really couldn't. My worries were mine alone and what they thought about me really didn't bother me as I had learned the skill to ignore others. It is no skill though, rather a behaviour you learn to adapt in certain situations.

I had time to think it over and over again, this was something that always happened to me, thinking subjects over and over and over again, but nothing out of the ordinary came into my mind, why everything was going bad for me and why was everyone, even teachers within my school turning against me, it seemed so hasty and unpredictable. It was the paranoia that was actually doing this, not actually the 'whole school.'

After the conversation with my parents, Adrienne must have been in contact with them, because they were well informed of what was happening at school. She and they must have arranged the whole thing with the school. I did have counselling before I met her, with a lovely Doctor, called Sarah. She was incredible and really helped me through my journey with depression, anxiety and paranoia. But at the time, it was not apparent I had these. She was really, well-trained and could pick up on these almost instantly, but that wasn't what I saw her for the first-time, I was having difficulties accepting who I was as a child and she helped me then, roughly around when I was 5 or 6, quite young to remember everything that happened!

I was so glad that the weekend came up as quick as it did as I certainly wasn't feeling happy in school with all the judging that was occurring and after the meeting. At least the weekends allowed me to stay at home with my family, a weekend that was away from a school which I hated and away from constant fear and away from reality. The weekends were the best time for me as I didn't have to go into school and could stay at home away from everyone else and away from everything. I was safe. I felt I couldn't trust anyone in the school and I even doubted trusting my family. I wasn't able to tell them certain things, like how I felt as I didn't want to sound paranoid, although I do strongly suggest you do tell your family any feelings you may be experiencing or thoughts you have. I guess that some people really don't bother to understand others and they should, the judgment of my current health was always being talked about and I couldn't cope with that. It was a lot better being at home rather than school, although I didn't go out, see anyone or do anything, it was better and I was free.

I just kept having the memories of primary school that kept coming up; I could see more flashbacks than I ever had before. I saw my old friends and me playing in the playground, all laughing and joking with each other, I couldn't believe that the times had gone forever. I still wasn't getting to grips of why primary school had gone and I thought I was. It was unreal for me to even think that I had got to grips but I thought I could hide that I had by not admitting it, but that didn't work. I wish I had just shouted out, I was scared and worried. But being me, I didn't and I kept this up.

I didn't want to speak to my family after what they had done with the school, so I sat in my room all weekend alone. I had many thoughts going around in my head and the memories that were close to me kept appearing. I couldn't live like this and being alone felt like the best thing for me, nobody knew me and I didn't know them either, but my family did, like I said I just didn't want them to worry for me.

The lonely weekend was over before it began, the time went by fast and I didn't want to go back. But it wasn't an option and it was yet another school day, this time it was more annoying than ever to return to that school. I just put on my brave face, walked into the school and ignored everyone. They couldn't see me in this state and they certainly couldn't know that I was scared. That is when I broke down again; I sat in the grand hall alone. I didn't want to know this school and that is when that lady appeared again…

Why was she there again? I didn't want her to help me; I didn't want to talk to her about myself. I didn't need her help, I didn't need anyone and I ran and ran until I eventually gave up and waited for her to catch me up. She wanted to evaluate my temperaments and whether I should be receiving extra help or be left my own way. It hurt to still think I was being judged and being spoken about that I eventually gave in and admitted that I was struggling. Never refuse genuine help, I urge you to think about who you tell your stories too, share your thoughts with and show emotion too, Adrienne was very trustworthy, it was my paranoia that made me feel she wasn't.

I think people at school thought I was losing the plot as I seemed short fused and very paranoid. By the time the first 2 years at secondary school had passed, I had told most of my class that I was still struggling to be happy in this school, I did like them and could tolerate them but it wasn't happiness that I was feeling, it was sadness. I had to try and see the good in everything I was always told, but when the good isn't there, how am I supposed to see it? I even didn't hang around with Melyssa, Ben, Kathryn and Mitchell, I think they didn't want to know me, they perhaps saw me as a different person to who they met at the beginning of year 7? Now being in Year 9, coming up-to year 10, I was very different, so maybe that's why they didn't bother? Melyssa was my next door neighbour funnily enough, she did have time for me. She used to check on me and come round to play board games or discuss pets. She tried anyhow.

I couldn't believe how most of them thought I was losing the plot, I hardly knew them, despite it almost being three years with them, and they hardly knew me and it wasn't as if I gave them my life story, they only really knew me because of my presence in the class. This thought was making me even more concerned, I was already worried that people were talking about me behind my back so it felt like another hit to me. Opinions will be opinions…

We all see what we want to see and if we all see negatively like I did then nobody will be seen as good and nothing will be good either, it will be like living in war times where nobody knew who to trust and nobody knew what was going to happen next. That time sure does sound a lot worse than this and the comparison may be higher, but can you imagine living in a negative world? Nobody with depression, anxiety or other mental health associations want to be seen negatively, they crave positivity, even if it doesn't seem like they do.

My recurring thoughts didn't all happen but the thoughts made me suffer setbacks and worry. I didn't have the time for anybody and I didn't have any best friends near my side to help or defend me, I was totally alone, well felt it anyway, and I wasn't going to be able to win this battle much longer by myself. I didn't know what was happening to me or what was going to happen to me.

Chapter 8, A turn for the worse

The 3rd year of secondary school was about to commence. My fears were still with me and my worries still hidden within. I told people that could see a change in me, that I was struggling but not that I was worried, concerned or full of fear. I really didn't want to be alive anymore, my life had become so pointless, I thought everyone was out to get me, be nasty to me and I couldn't put up with it any longer. I almost forgot to mention, that I had moved from my much loved flat, into a town house in St. Neots, I really didn't want this move, as I enjoyed being in the flat, we had lived there for just over 12 years and it felt home, home is where the heart is after-all. We had to move as my sister had been born and the flat was only two double bedrooms and she needed a room. School wasn't the only thing that was making me fearful, it was the new house too.

My heart beat was pacing faster and faster and I felt adrenaline inside of me and I felt hatred for myself. It really isn't a nice feeling hating yourself, or having such low confidence. I imagine many people experience these feelings, insecurities too are all involved in this feeling. It really became part of my life, my weight gain, acne and the way I was becoming all were contributing factors in my low confidence, as was the struggle I was having with the school and new house. My feelings got worse and continued getting worse as each day progressed, and I couldn't cope with my stress any longer, something really bad happened after this, something I had never attempted before and do not recommend. Because I was so low and unbeknown to me, depressed, I decided that it must happen, I must end this misery and that is when I swallowed all of my sleeping tablets and went to bed! I must have taken around 6 as the tablets were only collected from the pharmacy the previous week and I took two a night, I thought it would be the end of all of this pain and fear for me.

After taking these tablets, because I was in bed, I must have just zoned out. Nobody would know what had happened to me if I tried committing suicide quietly, it wasn't a big scene and I wanted to go quietly without causing worry to anyone, I always hated blood and seeing the way the press portrays suicide worried me, especially those who cut themselves, this really does concern me. They are all judged by the press and people throughout the country, which is wrong, because without the facts nobody ever knows how that people feels and what led them to try to end their lives, that is why judgment is so harsh and cruel.

I guess some judged me after taking me tablets, please don't judge me after reading this though, it took a lot of thought to include this into this chapter. Fortunately however, my suicide plan didn't work as I was woken up by my mother who was shaking me, she saw my tablet container had been emptied and I admitted to taking them. She must have kept shaking and shaking me, because she was very emotional after finding them and me just lying there in my bed. There were only 6 tablets left at the time which was very fortunate otherwise I may not be here now. My doctor had given me these tablets to be able to sleep at night, I wasn't able to rest as the constant thoughts I had, had taken control and the tablets seemed the easy way out. They did make me sleep easier, but never did they take away the worry.

The reasons behind these mood swings remained the same and I refused to change for anything. I couldn't just stop being unhappy for the sake of being happy, I couldn't change like a traffic light did. Change isn't easy when you have depression, people with depression do not see change as a good thing and often feel negativity, I did anyway and I really urge their families and friends to support them the best way they can. Don't leave them alone because they are low in mood, support them! I wanted to change and get by but I couldn't and I wasn't able to survive like this, this is why I tried ending my own life. I just couldn't change no matter how hard people tried to make me change I couldn't, I got aggressive and further paranoid, I was taking it out on family members after my attempt. I didn't want to be saved and I believed it was them that saved me. I remember throwing a cup of my mother and punching my father, nothing like this had ever happened. It wasn't the best thing to do but I was feeling so frustrated and felt that they betrayed me ultimately, I went into rage and it was all their fault that I was still having to put up with myself and my life. I would never dream of throwing a cup at my mother or punching my father now though, they've both helped me throughout my life and it just shows you, even the closest people in your life can be seen as strangers, the thoughts just took over and I took it out on them. This temper carried on building as did my lowness, I never experienced anything as bad as I did this.

My rages made it harder for me to even try being happy. The last time I went into school, I remember having an altercation with a person who wasn't even worth the time or day, this brought my fears alive- this school isn't for me. I got smacked around the face by this person who I hardly knew, I felt so much hate for this person after, acting like he was a big man. He told everyone about his plan to whack the boy with little confidence and little reason to live around the face. I remember this, because he had been in my science class, trying to patronize me and wind me up, with his pathetic remarks and jibes. I just let them fall. But when he came to attack me, after he told what seemed like most of the school, they all watched from both ends of the corridor, even the people I thought I could trust and thought was close to me. I didn't whack this boy back, I just let my emotions come out and that was the last proper day of school for me. I remember my English teacher, Mrs. Chillman coming down the corridor and witnessing this, she came rushing over to me, making sure I was okay. She was caring and I really wish to thank her for her care and support. She tried encouraging me, she certainly didn't allow bullying in any shape or form and I had been bullied. Attacked and broken, the school wasn't worth me attending anymore and I certainly wasn't going to attend after being violently attacked. I reckon the boy thought he was some kind of God but he really was a weak person for attacking a vulnerable and scared person. Even now I don't know why he did it, I think he just did it for attention, something similar happened to me, but without the violence. I had a drama performance and this girl was making harsh comments to me, trying to antagonize me, she thought she was clever and I lost my temper with her too, just like I did with my mother and father, so I chucked this object at her head, without even thinking, the reaction just came out. This object turned out to be metal and it just bounced from her face, luckily she wasn't blinded! But I never have done a thing like that again, you see when you're full of worry and not able to get out of the vicious circle, you change. I changed for sure. Why do people try to intimidate and bully vulnerable sufferers? I feel they obviously have no understanding.

As I wasn't attending school again, my thoughts worsened. I refused to go out and leave my house. I was in my room most of the time. I watched other people's lives on the television instead of living my own. Eastenders was a show I always watched, I recorded it, so I watched it over and over again. I still am a fan of it now and have been since I was a child, but this was my only enjoyment, seeing Dot in the launderette or Phil and Ian's rows, it wasn't my life, it was a soap! My studying wasn't at school, after these few rows happened. I did get suspended after throwing that object at the girl and the boy got suspended too, no charges came to either of us though, I didn't want to press charges.

The girl must have admitted she was patronizing me, because the school didn't suspend me for long. As I didn't work at school, my work was instead at home which allowed me to stay in all the time and not seeing the big 'bright world' didn't help I guess. Closing myself away from everybody else felt great, but now I can see that if I had let them in, perhaps things would be differently now? I wouldn't be so awkward or challenge everyone or everything before they have a chance? I didn't give a damn about that school, like I didn't when I first started and the feelings and hate for leaving my primary school that I once had must have been true. I wasn't this strong little child I once was, I was a weak teenager who didn't have any friends or any people to trust. I was so alone and hurt.

I had my emotions hidden and I didn't let anyone in, not even family this time! I told them I was fine and I wasn't, I kept telling them this, it has probably become apparent the more you read. That is my biggest regret not letting anyone else in. I didn't do the work that the school sent me, I couldn't understand it without tuition time and I just left the work without doing it, so much of was irrelevant to me, especially the Mathematics, I didn't want this rubbish. I even refused to do the Literacy work that had been sent home to me, Shakespeare and poems didn't appeal to me anymore. This is something that many people with mental health experience, their likes change and often they lose their hobbies and enjoyment out of life. I didn't even want to be in that school so why should I bother? I sat in my room with the work, I just had a sashed window to look out of and the world seemed like a rather daunting place for me to live in, especially out of that window. I didn't even have the contact with Josie much whilst I was suffering, in the later years whilst suffering I did though, and she was my closest adult friend. My family must have thought that I was better off alone. I didn't want to be alone deep down but I put the independent front on as I knew that this would make me appear more grown up and stronger. They did still check on me though, oh yes, it wasn't as if I was left alone in my room for the whole time I was there that would have been years! After I tried to end things, I think that is when they realized how much I was suffering. The time was going slower than what it did before, when I was on holiday having fun the time went past quick, but when I was alone and in pain the time went slower and it really had no positive effect on my low mood, doing nothing just going over memories can sometimes help, but all of memories felt long gone, with the wind, instead I just remembered the regrets I was gaining, they sure did gain over the years!

After my deep thoughts about being alone getting stronger, the school expected me to return to study full-time. I refused too- they didn't understand me, listen to me or even notice I was there, so why should I return? I wasn't going to go back at all especially after not receiving any help or support after I was attacked for no reason and had that altercation with that girl, it must be the talk of the school I thought! How dare they even expect me to return to a school that was like a living hell! I was studying from home and receiving work from the school which allowed me to be where I always was- at home, so why on earth would I want to return? The hatred inside of me for that school got deeper and worse every time I thought about it. I wasn't able to think happy thoughts and I was regretting even trusting them 'friends' of mine. Trust is an important factor in a relationship, without it, you can never feel totally able to warm up to a person or feel they are trustable. Therefore, friendships and relationships with lovers, would not work without trust. Not being able to trust anyone or anything is a big problem.

The four walls felt like they were closing in as each day went on and as a new day passed, another problem aroused. I could tell you every mark on my wall until it got painted. I was sick of the colour and painted it dark brown, I wasn't the best painter and it was a bit of a shock for my parents to see what I had done. Problems only made my life more challenging, as I expect they do everyone and at the time my problems ate me up. They had been doing this for some time. I should have thought positive but I never did! I couldn't see positive and nothing was positive to see. Telling yourself to feel positive and actually wanting to feel positive are completely different, it you cannot manage life you will not feel positive, so never assume a person with mental health can act positively, because this isn't fair on them, understanding must be had.

The walls appeared to be my new friends. It sounds weird I know, unless you have experienced the whole 'staying in your comfort zone' experience. I started talking to myself and writing notes down about what I wanted to happen after I died. I had even started arranging for my funeral, what kind of thing is that for a teenager to do? I could understand still why people thought I wasn't my normal self, I kept telling myself I was fine after all and quite clearly wasn't. I wasn't happy but I wasn't happy since I knew that I had left primary school so why should that make them think that I was different, there was a lot more behind my negativity, it wasn't just the school, losing friends, staying in or paranoia that caused me to become how I was, they were only minor factors in a major situation.

After all the problems I had with school and adjusting to new things that happened in life, I was taken to the family doctor in Eaton Socon, St. Neots at the time. I hardly went to the doctor and I didn't really want to go but it didn't matter what I wanted and I was made to go. I got told that I had to go to the doctors for a bad stomach ache I was having. I did have many bad stomach aches and that could have be caused to the constant worry, they were triggered more when I worried actually. Certainly not pleasant, but when worry takes over, I couldn't stop it.

After getting into the doctors, I believe I saw Dr. Mears, I hardly ever saw him as I saw Dr. Triggell, and she was my usual doctor, so perhaps she wasn't in on that day or had a book full of appointments? Who knows, but I didn't see her. Dr. Mears came out of his room and my mother and I stepped inside. I remember my Grandmother also came along, but she waited in the car, I think she was distraught because I had changed and quite dramatically too. After stepping inside his room I was asked tens of questions about my life, like: Do you enjoy life? Do you go out with friends? , Do you enjoy laughing? What has been the hardest thing you have had to do to date? Do you have any hobbies? Do you feel suicidal? Would you consider killing others? These questions seemed rather odd for a stomach ache cure I felt. Questions like this made me see that my life had been changed and something quite wasn't the same, I kept trying to tell myself I was fine, but knew really I wasn't fine. I thought it was something I had to experience in order to grow up, I wasn't aware of the consequences of ignoring my worsening ill-health. I didn't have enjoyment and was ignoring the fun needed in life that everybody should have, no matter what age they are, male or female, they deserve happiness! I wasn't enjoying anything and I didn't want friendships as I felt they all betrayed me eventually. I told the doctor that, when he asked me if I had any replies.

After answering all of the questions which the doctor asked me, the doctor hesitated and asked a further few questions, all were rather odd again and nothing related to stomach aches/pains, however it wasn't that what I was seeing Dr. Mears for, but for a diagnosis on something completely different. He diagnosed me with severe depression and I was put on Anti- Depressant tablets, I was told I had to take 4 a day; I guess everyone was right about having their concerns for me, even if some of them wasn't really helping me, they were right. It wasn't a nice feeling for someone of my age- 14 at the time, to be put on these anti-depressants as well as the sleeping tablets I had to take, just to get to sleep and some rest! It wasn't my fault the change happened. The doctor noticed that certain changes had contributed towards the depression and he said that I should take these tablets for the long run and that I should just try being happy, the tablets will help I reminiscence him telling me. I thought that his comment was the biggest insult to me ever. Being happy when you never see any good in anything is very hard and suffering from depression is somewhat harder. I felt I was being patronized, doesn't anyone understand that happiness cannot be switched on and off? Happiness cannot be brought or distributed, I had happiness but it had gone. Of course he must have known this, all of the years training to become a doctor takes.

"How can I be happy?" I asked the doctor. At that time I had blanked everything and everyone around me. Some people supported me better than others and some family members were so negative and unsupportive that I am surprised I talk to them now! They didn't have time for me, nor understand my battle with depression, because it hadn't been diagnosed at the time all of the incidents I mentioned to you occurred. I couldn't believe how unsupportive some people were to me. They really didn't care and they must have just thought that I was being nasty and conceited. I was agitated at the fact that they didn't even bother to care for me. Some family members did help which took away the ones that didn't although that was another betrayal, I had to get through this without them. Depression needs to be understood much more than it has been as some people only see the bad and it's the good they need to see. It has always been my aim to get everyone that has depression to be understood and that is why I decided to build my website.

The doctor stated that to be happy, you have to be able to see the good in things and people and enjoy life. All of these I was not able to do. Being happy is not as easy as some people think, money can't buy everything! Happiness comes with enjoyment, happy times good memories and I didn't have any of these nor any emotions. I wasn't able to see good in people, I saw the bad and that wasn't my decision it was just the way I saw things. I didn't decide to see all the bad that I did, it just happened. Thinking bad thoughts isn't good for you mentally or physically but I was unable to control them. I was unable to communicate with most because of my thoughts, I thought they all hated me and I was better off alone. Mother often communicates for me, she is my rock, without her a lot of what I wanted to say wouldn't have come out. I was more capable being alone but not capable of being happy than being in a room full of people. I needed to let people into my life and accept help. I am a strong-willed, determined, independent person who doesn't usually accept the help of others, but the 'strong-willed' I, certainly wasn't here in this part of my life.

After all of these questions, it rather felt like I was being questioned for a crime rather than a stomach ache, it suddenly clicked that my mother must have known I was experiencing depression, but she couldn't have diagnosed me, which was why I went to the doctors. Mothers can always sense these things, when a child changes and what help may be needed. Mines a star! I was given a prescription with the anti-depressant tablets on, I am aware now that they can change emotion and do have effects on people that some may say are negative, but in my honest opinion, the tablets do support you. But they are not the only thing needed to help you with depression.

I thought and thought over whether I should accept help from the doctor. I didn't want to get any more help from him, but after a forced discussion with family however, I asked the doctor to refer me to a shrink. He must have already suggested this to my mother when she telephone him to book this appointment, because both were in agreement with this idea. This was a difficult decision for me to make as I was beginning to admit to having the problem but I needed the help of a professional to help me through with it. Life had been like this for 3 years and I wanted to see the good, not the bad in everyone and everything. The doctor didn't even question me; he just looked at me and then referred me, that's what gave it away to me, that the appointment was for this and not what I first thought it had been about.

I finally admitted that I needed the help that I never thought I did. I wanted to become my old happy self and realized that to do this I needed to let go and be positive… it was going to be hard but I was ready to change. Admitting you need help can be a challenging thing for you to endure, look at my examples before I admitted I needed help, I could have asked for it a lot earlier, but didn't. I cannot completely answer why I didn't, but I can say one of the main reasons why- I felt I could cope with my emotion and by the time I realized I needed help, I didn't have much emotion left in me. Now, if you ever feel like I did or can relate your journey with mental health to this, please do admit if you need help. Nothings bad about asking for help, it is just over-coming the worry of admitting you have mental health that concerns most people. But, hand on heart, you needn't worry, it is quite common nowadays.

I got my tablets from the pharmacy and we drove back home, mother and Grandmother spoke quietly, obviously being informed about what had happened with the doctor, and I was advised my appointment would come through in a few weeks. Being a child, the service was a lot quicker and easier to access than what it is now as an adult. I tried to get an appointment to attend counselling again, as the depression has never really gone away and was advised of a long waiting list, it stuns me that hundreds of people with mental health are being put on long waiting lists, some are really urgent cases and get put to the top of the waiting list, but what about the others? Sometimes, if you were like me at the time, you may attempt to take your own life in a flash second, now how can you justify these long waiting lists? It is totally unacceptable and barbaric! More must be done, investment needs to be made for Mental Health, we're all the same and deserve good reliable treatment for our tough times waiting for appointments is stupid! I find it unbelievable that funding is being taken away from Mental Health and the shakeup of the NHS has resulted in Mental Health wards being taken away. Yet obesity is on the increase, like Mental Health and gets more investment. Mental Health and the psychological affects it has, can cause obesity due to comfort eating and many other factors, so surely there should be further investment from our government into both of these? It just doesn't make sense to me, that's my opinion anyway. I hope you understand it.

Chapter 9, Counselling Began

After the doctor's appointment, I sat and waited for my counselling appointment to come through, it came through in the post, quite quickly infact. Much quicker than the service offered now, with long waiting lists and not enough support. A sense of relief and worry arose when I received the appointment, but I knew I must go no matter what. In 2006, I had my first counselling session and I didn't find waiting for my new counsellor easy, I had never met her before but had been to the building, many of you may know it, The Newtown Clinic. I was becoming worried waiting, all thoughts came back into my mind, and I really hated this! I could hear somebody walking down the stairs. It must have been her as she came down to get me, she introduced herself as Dr. Pederson. She was an average height for a lady I guess, she had remarkable eyes and looked like she enjoyed holidaying as she was a lovely, olive colour. However after our introduction, I wasn't at ease talking to the professional about my problems. We walked upstairs to a patient room. She invited me to sit down, along with my mother who drove me back and forth to all of my appointments here. She said that we have to acknowledge the problems and try to fix them before anything can get better, we also needed to find the route of the problems I was experiencing and having, and it was going to take some time. This was a true point; I didn't want my life anymore and needed this help urgently. I told her everything that had contributed towards my fall such as the school transition, moving house, paranoia, weight problems and change of body which included my weight gain and acne. She took notes of everything I said, whilst acknowledging most of it with a nod and was very careful by what she said. She didn't let me know what she was writing down, this always remained unseen to me, but most of what she was writing down must have been what I said and what her thoughts were.

I did have a fear whilst waiting for her to appear of talking my worries and problems with a complete stranger, but like Adrienne before her, I had no idea who she was and she seemed to help me in some respect, although I was refusing to go to school, but worked from home, she was there for me and I hadn't known her previously, so talking to Dr. Pederson didn't seem all that bad, after the first meeting/appointment.

I remember after the first appointment, thinking she knew exactly what she was talking about and I can only thank her now for the time and effort she put into fixing me and helping me become less fearful. After the first session was over, an hour and a half I believe, I got a great deal of worry and anxiety out to her, she must have sensed that I kept a lot of my worry inside or only with close people I trusted, because she told me whatever I said would remain confidential, she wouldn't judge me and the problems I saw big, but perhaps some may see little, could all be addressed and acted on. It felt reassuring having heard this. We said our goodbyes and she said an appointment will be sent out in the post, which it was, for a fortnight or so later.

Before I knew it, time went by quick, despite only staying at home, having nothing to do. That's when the next appointment came through, it must have only been a few days, perhaps the next week before it arrived. My next counselling session saw things go from bad to worse, I cannot explain why, because I thought the first one went well considering I didn't know her and told her a great deal about me. I had to discuss all my problems again and not just the ones I wanted to tell her, all of them! Some more emotional than others. She must have known that the problems lied deeper inside of me than I was revealing to her and she was right, she must be able to have read me. The fact that I was talking these problems over was hard but then it got easier with time as I knew my counsellor longer. It was still a very hard time and my early teenage years were not enjoyed or much fun had but full of hate. You hear about how the teenage years are the greatest, unfortunately mine didn't feel like that, some years were better than others, but I wish I could have changed how I thought, but I couldn't. I told the counsellor about my life but struggled to acknowledge why I didn't trust people? The main problem I was facing was the risk of losing everyone I loved or cared for... just because I had depression it didn't mean that I forgot them, I just didn't trust all of them and my trust was probably not something they all cared for anyway. Some left me to suffer in silence and the supportive ones stayed and helped. It showed me who my true friends were and their true colours for sure!

The trust between me and my aid built over the months I had counselling. I felt like I was getting somewhere with my counsellor, we had an mutual understanding of what had gone wrong for me and what had happened to make it go wrong. It was hard for me to go especially when I was being judged by some family members for having counselling, they all deny it now though, but they did judge me for not going to school and getting the most out of my education, but how on earth do they know how I was feeling, when they've never had depression? It does become irritating when you hear people judge others, they aren't perfect either remember! Just keep that thought in your head. I just kept my brave face on and thought I will fight this condition and I will win, there was beginning to be a little fight back in me! I wasn't going to let this take control of my life any longer, I had enough of it- I wanted my old life back! Secondary school wasn't for me as I have admitted to you, some of my friends and family brought me pain and I feared of losing what I had left, the ones I had left anyway. I couldn't trust everyone, yet I had a worry about them. I had a big fear that I would lose them. If I did lose them, they would never come back to me and I didn't want that to happen, the people worth having in life will always stay with you no matter what and the people that don't are not worth worrying over, yes it is hard to move on from people you want in your life, I would know because my eyes were opened to this in Year Six and in the years I attended my secondary school, but ultimately, you cannot force somebody to be around for you, if they do not want to be. That's reality!

Once we, my counsellor and I both addressed my problems, my aid did her best to try and come up with ideas and strategies to help me both physically and mentally. My characteristics had slumped, I wasn't happy with being me; I hated my body and couldn't look in the mirror! It is surprising how little confidence I had, now I still suffer with confidence problems, but put up fake confidence to show others that I am confident, although deep down I am not really a confident person, infact shy. I was ashamed of everything I had become and what I had become wasn't what I wanted to become and that feeling was another one of the worst feelings I ever remember having. I didn't want to become a person that was scared to go out, too frightened of life and tried to end his life through over-dosing on sleeping tablets, I wanted to be a person who enjoyed life and had great trustworthy friends without the whole trust issue.

I always spoke my mind and didn't care about other people's views or feelings. When people pushed me too far, which some tried, I lost control of myself and said some hurtful things. I would say I regret them, but as previously mentioned, if you provoke a person with mental health for own amusement, then that person provoking isn't worthy of an apology. I had a motto which I used quite a lot- don't like it, lump it! It kept my opinion about others intact. I guess people saw me as a rude and quite opinionated but their opinions didn't matter to me, all that mattered was me and working out ways on how to try to get better. I was lucky really that my mother had spotted that I had been changing in temperament otherwise I may have got really worse and not been as able to fight depression as I was. It never does go away though, different days bring new challenges, people are challenging and life as a whole is challenging, but what counts is being able to wake up every morning and at least trying, because it's the not trying that begins to make things worse.

After a few successful counselling sessions, each being rather different to what I had expected. Which was to be challenged at every opportunity and judged, I thought that things were getting better for me, as I was beginning to admit my fears and problems with Dr. Pederson, but they were not! In fact it was going the opposite way again or seemed like that anyway. I tried taking a giant step forward but I was taking 2 steps back. Sometimes you have to change before you become better as a person and the change was the hardest thing to commit to despite how hard I tried and being a trier, I sure did try hard.

The third session was the most serious of them all; it was only Dr. Pederson and I, mother was asked to stay outside the room, probably to build on strict confidence between the two of us. I had to say who I could trust and who I couldn't trust and give reasons why and why I felt like this about them. The trust list was emptier than the list of people that I couldn't trust. It was hard for me to say that I didn't trust close friends and family because I was worried that they was plotting against me behind my back, part of a grand plan, of course it was because of anxiety that made me think like this. They left me when I needed them the most and just thought I was challenging, I needed the help they didn't give, and it always hurts me to think of that! I had a lot of altercations with my family and most of them all thought that is was because I was becoming a teenager that was the 'problem'. They didn't see that the depression had taken over me; some even thought I was being vile and changed the way they were around me, it made me uncomfortable. Again their points didn't knock me, they just sickened me. My own family thought that about me? What do others think about me? I think that the trust issue laid within my friend group more than my family even though they had negative comments, some did anyway... The secondary school group of 'friends' however did betray my trust as they didn't support me through my hurtful time at school; they didn't even help me when I got hit. The trust was hard. They instead, stood and watch me being attacked, it was unforgiveable and unforgettable!

I mentioned this particular problem with my counsellor and she said that the negative thoughts of any person could change what another person thinks about them, it gives the other person an option to see what the other sees and sometimes we all judge a book by its cover and it's like that with people. I think that my negativity was triggered by opinions. I stepped down from being as opinionated, especially as I hardly spoke much then, and I just allowed people to say what they wanted. I wasn't going to argue or express hatred with them now, they weren't worth it anyway, and I had enough to deal with, without worrying more and more about who disliked me for no apparent reason. The thing that hurt the most was that I couldn't trust. I just wanted to be' happy', who doesn't? And as I took a few steps back, my dosage of anti-depressants was increased. I originally had to take around 4 tablets a day just to stop me from being so negative, this was then increased to 6 tablets a day, two in the morning, two at lunch and 2 before bed. Things really did take a tragic turn again and I get really upset telling you this, but I tried committing suicide again, meaning I tried twice, I wrote a letter to my family and said I was sorry but it had to be done, I don't understand this myself, why I thought because of what happened I had to stop my existence on earth, I really felt this way at the time I guess. And again, I want to express the understanding that suicide is never a problem solver, most fears can be dealt with, with time, so never try. I overdosed on the anti-depressant tablets this time, rather than the sleeping-tablets again and dropped the container on the floor. I didn't fall asleep, I collapsed on my floor for a while then I remember hearing my mother shouting and calling the NHS Direct line and my counsellor. It was a second experience of overdosing but again I pulled through, somebody from the heavens must have been watching me. At the time, suicide felt like the right option again, but now looking back at both attempts, I am glad that I didn't die, I don't want to die! It was very fortunate my mother had finished work after a long day of working, if she hadn't been at home in the evenings with me, then who knows what could happen? She began taking control of my tablets after this happened, it probably was best she did.

The doctor on the telephone gave guidance to both me and my mother and he instructed my mother on what to do for me. I had to be given my medication by her as I wasn't trusted to take it myself. I was watched to make sure that I was taking the correct dosage and not over-dosing myself. This alone was a hard thing to do as I had to accept another person dictating to me what I do. I really wanted to take the medication myself as somebody else giving me my tablets felt like I was a child again, being told what to do really wasn't appreciated or wanted by me, but I had to be, I wasn't in a good state.

Once my counsellor found out what had happened on the telephone and what I had tried to do, I was booked in for ten further sessions with her immediately and we had to discuss my suicide plot. This was very emotional and hard. I didn't know why I wanted to take my own life, I just couldn't explain why. It wasn't selfishness, I didn't think of my family when I tried though, but thought rather of myself. I just thought of the easy way out again I guess. That way I would go and my family would be happy without me, little did I know that they would actually miss me. It was odd coming to think about this, that I didn't consider them? Now I know that they would miss me and it wouldn't be fair on them if I did that, they all would be distraught and wrecked ….I would be the selfish one! I never thought they noticed me, I was just in my room without anyone to talk to most of the time, I felt so alone that the reason behind me wasn't worth living, but I was worthy of live and that's why my attempts to end my life didn't succeed, this must be the reason. I had this imagination of them finding me after I attempted suicide, I thought I would never want to find somebody else like that so I wouldn't expect them to find me like that. But it is all very well me telling you this, but depression affects the brain significantly, there's nothing the same about a person and it has many different triggers. Low-self-esteem is a big factor with depression and I was feeling really low in this.

Talking my mind over with my counsellor was tricky and intense; it was like reading somebody else's brain as I didn't know what was going on in mine anymore, let alone hers. Somebody else was inside of me I remember thinking, 10% me, 90 % somebody else. My life had changed so much compared to the prior years. It was very different and more complicated than ever before. I was struggling, more than I ever struggled before. I was struggling more than I ever did and wasn't getting any better even with the tablets, they just relieved some of the pain I was going through at the time. They couldn't erase my memories or fears; they couldn't even make me happy when I took them or change who I was at this time in my life.

The thoughts I remember having were of an argument I had, I kept going through my head to think about it. My thoughts were mainly based on that argument that I had with my friend group within school about my opinion. My opinions weren't fair and we had a row and they all fell out with me. I didn't want them to do something because it made them happy and didn't make me happy. I was very mean and bitter towards them. I guess that was the reason they didn't care when I got attacked, I resented their happiness as I didn't have any. Voices appeared in my head telling me to ignore the argument but for the first time, I followed what I wanted to do without judging or miscalculating my mind. I wanted to make up with them and rejoin their group even if I didn't see them all of the time which made it hard to tell them. I had enough of my hate and didn't tell them how I really felt….they didn't even know I suffered from depression and just thought my halo had broken! It really does pay not to say anything at all if you cannot say anything nice. I learned this and just wish I didn't say anything at the time but I cannot change the past, I can only try. I lost all the friends I ever liked or needed, even if they went against me in the end it was because of what I did to them before. I guess that I should have tried to make up and apologies to them before it all happened and school but I didn't and the loss of more friends was on my mind for some time.

Another thought I had was about the move I had to the town-house, I didn't really want to move and this affected me. Having Melyssa next door though was a good thing as she did check on me and must have been concerned when I never attended secondary school much.

Dr. Pederson thought it would be good if I did a plan about what I should do if I have anything to say to people and how to address my fears in a structured sense. I didn't have any structure about how I could express my concerns or opinions to people. The only way I did this was by saying it out loud to people without thinking about their feelings. I wasn't one to bite my tongue before I spoke and this did make it hard, even when I was quiet I could still get into arguments. I was still not being positive, I didn't want to change myself for others….it was my life after all and I should live it how I want to I always thought. It began to feel like that although I wanted to try to fight depression at one stage, the next stage I would have given up. In that session we also created a plan on negative thoughts. We looked behind the reasons that could be making me more depressed and the following made it harder- I couldn't sleep some nights at all despite the sleeping tablets, because I would lie awake worrying or I was kept awake because of problems which are addressed later on in my book. I had a very short-temper and often took what people were trying to do and say the wrong way, especially as I had trust issues. I had many worries and these worries had taken over my life, who I was and my good traits. She thought it would be best for me to write any of them, my problems, down before I attempted to go to sleep as this could help take the worry away.

My main worries were- The House, Friends and Family, Betrayal, Death and Failing.

I wanted to try and forget about my worries so I wrote them down before I tried going to sleep and this didn't help straight away, I did keep trying to do this though. So I was wide awake some nights just worrying. Not being able to sleep does make you less concentrated, more provoked and very moody, it's not healthy either. The night's I lied awake led into days I was awake, I couldn't switch off my worrying and it wasn't made easier due to people. I could go hours without sleeping because of worry.

I carried on speaking my mind to people and destroying some strong relationships which I had in the progress, in particular with the few friends I had left. I didn't see them, but spoke to them on the phone or through text message. I was very grouchy and not being able to sleep was a main cause for this, as well as my anxiety, depression and paranoia. I wasn't expecting miracles overnight but all I wanted was a good night sleep and to at least try and forget my worries for one night, at least!

I think I worried about being betrayed because I had been betrayed so many times in the past, I betrayed my primary school friends by leaving them behind and my secondary school friends betrayed me. It is kind of like a karma I kept telling myself. I told this friend I didn't like much about my life, I even told them I had depression and this friend was very close to me from secondary school, someone I actually had classed as a friend in that school! I never thought I would have close friends from secondary school, but I had this one, at least I thought I did, he was in that group of 4 which I mentioned. I knew my instincts were right before I even started secondary school and I should have listened to them better. That friend ruined our friendship because that friend was in the group of friends who betrayed me in the final weeks of me attending school and soon after me leaving school, the news about my depression spread to some other people in the school that added their own stamp on my downfall- it didn't help at all, but whilst suffering from depression, I shut the door on everything I cared about, I didn't need friends and I certainly didn't care who knew about me suffering as I was adamant that I would beat depression. I got messages some caring, some not so caring or nice! I couldn't cope with another problem.

I also was worried about death because it's something that we cannot stop and because I lost family members to it, some of whom I really miss. It certainly is not a nice subject to talk about. This fear was also with me most nights and a few times I could see shadows and heard people talking, I was really scared and petrified, I feared going to sleep in case they got me and took me away. That worry kept me awake, along with other worries and it was like I was going slightly weird because I couldn't see the person, just shadows and voices- it wasn't logical. I couldn't understand why only I was seeing these shadows and I feared the worst like I usually do. I must have been around 14-15 years old at this time and I remember this feeling well, I have experienced a similar encounter with hearing voices a few times since this, once when I was 18 and recently in my 20's. I really cannot explain why it happens, but it really is mystifying.

As well as the fears above, another big fear of mine was failing, this was a big fear of mine, especially as I had not been attending school and was not having enough 'in-school time' to be likely to pass my GCSES. When my mother went up to that school to discuss me and my downfall, I remember her heated argument with the headmaster, he was very rude and she wasn't having any of his rudeness. He, being a headmaster, didn't think the school had done anything wrong, he thought the blame was all on me. One of his assistant principals even told me mother "Your son will fail and be good at nothing if he doesn't do his GCSES!"

What an absolute disgrace, she was furious and sure told them what for, I remember her coming home from this meeting and telling me what had happened, despite the school providing Adrienne to support me, they never offered support except through her. No understanding of depression, anxiety or mental health was had and it is shocking. I can see in this age we live, that many schools and staff have lacked knowledge of mental health and do not really understand it, I often say to experience something you have to go through it, no amount of 'textbooks' can inform somebody what it is like to have depression and mental health, you have to experience it to know. And of course, this headmaster and his staff did not experience it, they were self-righteous fools and I recently (2012) contacted my local newspaper and offered them my story of my experiences. I was surprised with the replies I got back, one was from somebody I knew called Lisa. Lisa and I attended a childhood care club together, for some years when we were younger. She agreed strongly with what I put in the article and replied to show that I wasn't alone and she was glad she wasn't alone either. It goes to show, there is not enough support in all schools for mental health or bullying and it certainly is often underappreciated, I can say that with great confidence. Staff in schools need further training on mental health, don't get me wrong, some may be extremely good at supporting, like Adrienne was, but she was trained in this and so many are not and think they can support, when in actual fact they cannot.

As you can imagine, the meeting was not a success and even now, when the school replied to my article they stated that they did everything they could for me and would welcome feedback, well if they cannot see the obvious, it isn't worth helping them!

Anyway, back to my fear of failing. I hated failing and making mistakes because I worried that they would return to destroy me and failing was never a good thing anyway or at least I thought. I just thought what if that was a mistake and that mistake then turned into ten mistakes and then the amount of mistakes I made increased and the worrying of failure was very confusing and nothing I could do could change my worry. Becoming a failure was never an option for me despite having depression, as long as you stay strong and try your best to fight it with support from your close loved ones you will win and be a winner! I guess some people saw me as a failure whilst suffering from depression, but that was ok as I did fail in some things and I admit I failed. Failing isn't a bad thing really, because if you want something so much or badly, you will keep getting up and trying for it. Depression isn't a failing, it has taught me so much, living with it has been hard but I wouldn't be who I was today, had I not had it and continue to live with it.

I decided that I should follow Dr. Pederson's advice and wrote down these fears, I was experiencing and what happened when I felt them. It was like writing a log of recent events, in one case I remember writing about feeling another cold presence and getting a voice in my head that told me to forget. I didn't understand what it meant and didn't believe it was real. It was like the time I could see shadows and heard them talking. Could it be a message from beyond the grave or just what I wanted to hear? This question still is unanswered because my counsellor didn't know how to acknowledge this. She said that sometimes we have voices in our heads that tell us what we want to hear, I did want to forget my depression but I couldn't because I suffered from it and it had/has become part of my daily life, and still hadn't had enough help to see positivity or the clearer picture of what I wanted to be. It's not always the easy route that helps you succeed, it can be a long route back and forth many of times to succeed.

Writing the notes down did help to an extent, I knew I had them written down, but at the same time I couldn't stop thinking about what I had written down. I did keep a few of these notes and took them to counselling when I went every week. I can see a note I wrote now and it read:

"Note 10, they keep entering my head, telling me stuff. There is many voices here. They are worrying me. Some are telling me to ignore it, but I cannot ignore it. The other one said to me, you need to give up and let it be. I couldn't make sense of it at all."

Note 10 still confuses me so much now. I had so much ill feeling towards what was left of my friends and family that it began to hurt. Something I was feeling hadn't happened for a while. I could feel for others! This was a major relief, although not caring did seem a lot easier.... I could see what others wanted which I hadn't been able to see or feel in sometime. It felt so surreal and like a small weight had eventually been lifted from me. One weight lifted from me, didn't help everything though as I still had many more weights on me! I let my counsellor know that I could feel what others felt and that I was thinking more about them and not worrying so much about myself. I wasn't sure whether this was a good thing or bad as the bad feeling still was there and it didn't seem to be getting any easier. Thinking of others did take away my worry of myself and my fears, I do this now, I try to help everyone else to take my worries away from me, I cannot help it. At college, I studied business, we had a group of like eleven males and two females, and I recall helping at least seven of that group with understanding, assignments and peer-support. It took away the worry of me and it did help whilst doing it. I do suggest, if you feel you can, trying to help others, but of course it isn't always easy, it does depend on your mood and how you feel on particular days- each day is different.

Feeling emotions again and sensing what others felt was a turning point, a u turn in my life that helped with some areas of my depression. It didn't help with all of my depression though as it was more severe than that, as I still wasn't trusting people or seeing positive and hardly set foot outside my four walls. I did occasionally go out to socialize with a neighbour I had befriended down my road, but that was about it. I fortunately wasn't thinking of suicide but I was still on my high dosage of anti-depressants, as well as my sleeping tablets. These were helping me somewhat, but not completely. Medication doesn't solve all of the problems, despite the thousands of them there are around! I felt like a prisoner in my own home, I wouldn't go out a great deal and most of the sunlight I saw was out of my bedroom window. I hated even contemplating looking in front of a mirror because when I did I didn't like what I saw, my stomach was horrible to see as was all of the acne, it must have been puberty as I was that age, but what I had trouble with most was my confidence when approaching the mirror. I didn't like what I saw. Sometimes it felt like I was a stranger just looking in at my life and I couldn't change. I was gaining more weight so I starved myself and stopped eating meat products, my skin then went really white and I wasn't physically active, researchers now say exercise helps with depression, had I known that at the time, I would have tried, instead of just sitting around. Being physical does require energy and I didn't have this, because I wasn't positive. I do feel that positivity is a major part of exercise, because you need motivation and commitment to do it. So not everyone with depression will feel able to exercise on a daily basis, I certainly didn't.

During our teenage years, change does happen and our bodies change, acne also decreases confidence as does putting on weight, but it is a healthy part of life, the change from child-teenager-young adult. I hated putting on more weight and stepping on the scales as I knew that I was getting bigger and I couldn't fight it as I refused to eat, I didn't eat for 4 days in a row at one time. It wasn't healthy to do this and my body suffered further due to my skin becoming pale white and weak. I didn't lose any weight during this time though, no exercise wasn't completed and a healthy diet and exercise is needed apparently to lose weight. My confidence was at an all-time low.

I just wanted to focus on my life, but I couldn't. As I didn't go out, I didn't have any relationships with anyone and my socializing wasn't really much to talk about. I did have a friend down my road who I occasionally spoke with. Everyone else spoke about their outgoings and how much they enjoyed them and I just spoke about my worries. I wasn't like them, I didn't have the fun and enjoyment out of my life as they were having out of theirs and it did hurt me,

I felt jealous of them. I did want to go out and enjoy time with my friends and family but something was stopping me…it wasn't fair and I had not control or knowledge on how to stop it. What is often misunderstood is that we cannot, commit to change just like that. Depression isn't easily cured and some doubt that it actually leaves those suffering from it, I agree with this. Every day depression can arise in different situations, I have had it since 14 years of age and it certainly can be unexpected, those who don't have it can be critical and unless they live with it, have no right to criticize. Now, at 20, I do still have it, but some days are easier than others, and some days are much harder than others. So it can be questioned whether depression can actually be cured? But with understanding and support, it can be helped for sure, I know this as a fact!

Having fears and trust issues made it difficult. I couldn't trust anyone who tried to become my friend due to my previous experiences so those who tried to become my friends didn't, although hardly anyone tried face to face as I didn't leave the house much, they would try on social networking as they knew me through word of mouth in the school. I hardly saw anyone anyway and the only time I would see people was when I visited a few family members or attended my clinic appointments. So I couldn't build friendships as well as the leisurely teenagers do. It did have a damaging effect on my confident further.

The times I tried to do things and enjoy myself are beyond countable. I couldn't go out because I was worried and didn't want people to see me when I was a de-motivated, unconfident and sad person. This felt hard for me to do, I hated me as a whole and didn't want to show this. I tried never to drag people down with my problems, although sometimes I guess they had to put up with me being like that, I didn't want them to be involved; the depression was my problem and I thought the best way to win was to fight it alone, despite the help I received I did still feel alone. I didn't have my counsellor with me 24/7 so it wasn't as if she could see my mind or live my life, she could just assist with the information and problems I told her and give her well-educated opinions. I felt really bad after the altercations I had with my parents that I doubted they really would forgive me, but they did. They knew I wasn't really thinking, it was fortunate.

Being alone, well feeling alone, wasn't a nice feeling especially if you have family. Families are supposed to help with problems but my problem seemed too big for most to understand or help me with. Perhaps I did speak my mind a little too much, but they didn't understand that it wasn't me speaking and that I wasn't on planet- Cheshire cat; I was on a what can be described as a 'different planet' to them, planet unhappy and depressed. I couldn't help it if I felt this way, everything I was had gone, everything I believed in and wanted disappeared, so being happy was never going to be as easy as they thought.. I still had no confidence and my life felt like I was swimming in the deep end. I couldn't beat depression alone despite thinking I could and no matter how hard anyone tells you to do something, if you don't really want to do it, you will not. Besides, somebody telling you "stop being like that" when depressed is hardly any support! You're better off without it.

As I openly tell you this, this is how it had become with my counsellor. I tried my best to reveal everything to her, it would be easier if I told her everything and not only what I wanted her to hear, otherwise how can the help given be genuine? You must always be honest and express your thoughts/feelings how they are, never hide them or cover them, because you worry about others. Be clear, be you!

I attended my next session of counselling and expressed to her that I was worried about attending because of what people were saying about me. As the news had got to more and more people about my health, some people started calling me "mad" and this did deeply hurt. I didn't want to be called mad so didn't go to see my aid. Again I did a wrong thing as her help would be the help I needed most. This wasn't the right decision and I regretted missing the session. Judgment, again came in to this, people were shouting at me as if I wasn't human, all I was doing was entering a clinic and they'd stand at the roadside shouting this, I didn't even know them, but they were in a group, so I just walked past them, ignoring their offensive comments, as did mother. She didn't have time for these 'time-wasters'. As well as this, the news was spreading around the school, because mother had a few telephone from teachers asking how I was and one even told her what they had heard, it turns out that someone from that group, was a student in the school and had told someone who had told somebody else and it spread. It was my business, it should have stayed like that.

Understanding people is a challenging thing and some people are mean and very cruel. I often say a person has their reasons for why they are like they are. You can listen to their views but you don't have to change because of them. I remember my counsellor agreeing to that statement of mine and saying, "Everyone should be who they want to be, obviously you are not happy and the depression is taking your motivation and enjoyment, but we can help you sort this, even if you get a bit more satisfied, it is better than not at all." She then moved onto saying that bullies are people that have been bullied before and they think they are the bigger person because if allowed, they can ruin lives. Bullying is a huge factor in society, as is discrimination, yet many of it goes unreported, I really feel it needs tackling and standing up against! Bullying disgusts me, I understand the need to express anger after being provoked, but never should bullying occur because of indifference. This isn't what we are on this earth for, difference occurs, live with it!

After hearing that I could get help with my depressive state of mind, I was very glad to find the help she offered and decided to return, despite the shouting I had endured previously from 'fools'. I knew that I wasn't the only person in the world to suffer from depression/mental health, of course I wasn't and she must have had experience of previous patients with the condition as well. When I entered the waiting room to wait for her, there were other people waiting for their appointments too, so of course I wasn't the only person. It gave me assurance that others have depression, now through my website, I try to support depression, because I live with it and have experience and guidance/support that I hope can help others. She had answers to all of my worries, well most of them anyway, and all different ways on how to try and fix my fears that I couldn't believe, writing them down for instance. Perhaps this was a simple tactic, but I didn't think of it, would you have?

As I was now able to feel others and become more affected by what they said, I decided whether I listened to them or not, sometimes I did and sometimes I did not. It depended on what they had to say really. I didn't want to hear their negatives because I couldn't cope with more negatives and I wouldn't see positive, I still struggle with this and am questioned why I question reality in different situations. I wanted to hear positive comments to help me see positively and not negatively! A good mixture of positive and negatives thoughts are needed to be happy I guess, constant negativity all of the time is enough to scar a person for life.

I wasn't the best at mixing positive and negatives at the time and this would be the next thing that my counsellor and I would try to work on and try to sort. Everything takes time though, remember that and never try rushing, because it can cause upset down the line of your attempts to get better.

We both came up with a spider-diagram, this would enable me to put in my positives and negative thoughts, and then each session and whilst at home alone. I would put down what has caused the positive thought and draw branches from it, then the same with negative, what the thought was and how it was caused. This did help, because it enabled her to see what problems I was having or what made me feel a little happier, by looking at a graph.

Chapter 10, The Other Factors

I continued to do these spider diagrams for quite a few months, and it became clear to her, Dr. Pederson, that secondary school wasn't not my only problem or major concern. Oh no, it wasn't just the school, although that and this other worry would have probably been half/half, 50/50, both causing the worry. I mentioned the move from my much loved flat earlier, now you'll see what the problem with this was.

To try to sort all of my problems out though, we needed time to see why I wasn't seeing any positive thoughts or thinking positive. One of the main reasons that I wasn't happy was because of the change. Not just the change that occurred when leaving primary school, the change that happened after departing primary school and joining secondary school. That was a big worry for me, but it wasn't as bad as time passed, being in that secondary school was a big worry of mine, I never did feel confident there, nor did it become easier. But this worry wasn't about school, it did happen after primary school transition though, it was a move in property. The move from my family home to an awful home that I classed as worse than hell.

Having a family of four, a two bedroom flat was not big enough for us all, there wasn't any room for my sister once she was born, so my parents decided that we needed to move, so we did move and into a town-house. Which was quite nice to look at, it wasn't overly sized, but had enough room in for us four, big enough for our needs. The area I lived in was an area which wasn't good for my health at all. It was great when we first moved in, there was a nicely landscaped square which had flowers in, the roads were kept clean and the neighbours all got along. Melyssa was my next door neighbour and her family was lovely. Her mother and mine became friends and it was good for that to happen. Melyssa and I went to school together, when we first moved in. I must have been around 13, as must she have been. This was when I went to school, I did try to go even after not wanting to go after the transition, and my attendance was much higher than it got in later years for sure. The area itself, being close to town, was surprisingly quiet and full of what seemed like neighbourhood spirit. But all of this was about to change and within a year or two, the area suddenly had become full of violence, drugs and crime and it wasn't my nature to enjoy this atmosphere, nor was it any of the other neighbours, I expect. Why would I want to put up with all of these extra problems, when my problems alone were causing me heartache? It was a living-hell for sure, no sleep despite having sleeping tablets, and the youths would hang outside until early morning, having nothing to do but cause loudness and nuisance to the properties down that road. It was destroying the community we had down that road, it did succeed in some sense.

When we first moved, it was something to look forward too. A new house, new things and new neighbours but leaving the neighbours that we already had behind was hard and although I wasn't able to see that being a problem it was a huge problem as I lost their friendship also, I didn't do well at keeping friendships when I didn't see a person and they all seemed to be left behind of me, that bothered me just like leaving my old friends behind did. Why did it happen? I kept questioning myself, but it must have been because we needed to move and not everyone down our old road had time to constantly keep in touch, overtime long distance relationships do fail, especially if no effort is made on either part. This must be why. The area was a nice area when we looked around the property and we were told about all the 'nice' things about it, there were nice surroundings, the house was close to the town centre but not to close and close to fields where you could take a walk by yourself or with your dog. It was actually quite nice and we did settle in to the house, it didn't seem like our flat did, in the sense of a homely feel, but it would be our new house and home, so that would come, well we thought it would.

But within being in our new house for only 6 months, the area somehow managed to turn from being a good area into a dispersal area where the police visited three times a day! It was terrible! Never had we lived in an area like this, our flat was lovely and peaceful, we never experienced any problems there, the neighbours all were lovely and this would never have been accepted in that road.

Of course, I didn't like where we lived after this negative impact took effect, the area was getting worse and worse as the hours ticked and if I couldn't be happy at home, I couldn't be happy anywhere! Considering I was shut in my room for most of my time in that area and not attending school, it was hard to deal with my problems and the problems that I was getting in that area, the problems all added up and all my problems went together, more problems for me. My life was becoming a drag and negative thoughts took control of everything I did within my home. The house was a big problem for me, as I didn't want to be in it. It wasn't the same house I knew or the same house which I grown up in. I just wanted everything back to the way it was, but the past isn't the future and it wouldn't have been possible for this to happen either. Unfortunately for me, Melyssa and her family would move shortly and they didn't move close either, they moved quite a distance away. It was upsetting that Melyssa moved, because she never knew that I classed her as a close companion. We often baked together, we wasn't all that good, nor would be worthy of winning Master Chef, but we had enjoyment and made Ginger Bread Men, although I recall we did burn a few, but it was fun. Fun is what I needed! Laughter is good for the soul and mood you see. Her moving did impact me, even now thinking of a close friend moving, does hurt. I hardly see her now, we do occasionally keep in touch, but it is never the same as having a person around you, on a daily basis, is it?

I couldn't look out of my own bedroom window without worrying or without the youths swearing at me for doing so. It felt as if they were constantly around, looking for me. I had so much hatred for this and had another negative feeling, the feeling was as if I was being watched all the time. I'm not sure if the paranoia, depression or anxiety I had triggered this feeling, but I was so worried and at unease thinking of it. I had every right to look out of my window, the window where I saw everything going on rather than actually being involved in it myself, it was hard not to look out of the window after the amount of trouble they caused. I did eventually stop looking out of my window as it wasn't worth keep getting abuse for looking, so I had my curtains shut most of the time, so I wasn't seen and after the pathetic idiotic behavior that kept taking place, I couldn't see my life even getting better, but I wasn't going to commit suicide or try attempting it again, I felt more able to cope this time, I didn't even think about doing so anyway, which I guess was a good thing, when suicide is approached in schools, what I would suggest more is explained of is the thoughts and feelings that person was going through rather than they simply committed suicide. Like anything in life, we all have our limits which we are pushed towards, so this must be identified, it's a lot more complex than just ending life. I didn't have scars on the outside from trying to overdose, twice, on my tablets, but I had internal wounds that did really hurt me to think about, even now I consider whether I should have done it or not, I do have mixed opinions.

As I wasn't at school, I couldn't see anyone from the school anymore and my personal life wasn't really worth talking about. I had the trust issues with them from the school anyway, so wouldn't really want to see them, I couldn't either as I didn't go out. With Melyssa gone, the days seemed much longer, I didn't have her popping in or talking to me over the garden fence, and those days were long gone, unfortunately.

It seemed that whenever I did go out, for a counselling session or with my mother for her errands, that the bullying from the congregating youths was constant, I hated bullies then and I hate bullies now, the bullies really are not worth worrying over but when they are constantly on at you it's hard to forget or accept. Bullying should never be accepted or tolerated but I put up with it although deep down I was broken. I did fight back a few times though, I had been wound up by their jibes and nasty comments, they knew that I was experiencing a rough time, so one of them thought making a death threat was pleasant, my family and I wasn't amused, so my parents had words with this unemployed youth and made sure he wouldn't cause me trouble again. This did work for a while, but it didn't stop them coming round the area, causing havoc, loudness and downgrading the area through vandalism and junk. Bullying is on the increase, even now after so much is known about it, it happens in the workplace, school, out of the workplace and school, so why do people bully? I'll leave that up-to you to decide, because my answers may be very different to yours.

The bullying was not even happening for a reason, it was happening because I was different to them, I wasn't the loudest person around and I was shy and fearful. I guess they saw this as I didn't go out much and they must have picked up on this. I didn't want to put up with any more problems from them because the abuse they shouted up at my window was more than enough, I wasn't the only one they caused problems for though, they smashed my neighbour's window, never have I experienced such a bad case of behaviour and it just truly shows, that without anything to do or work to commit towards, the youths will be misled into crime and patronizing people. Perhaps the government could create a scheme that works? Other than the NEET scheme, which has only helped half of the unemployed and not trained young? I would have been able to ignore this before but I could now feel for others and I could feel my own personality, I was caring and a trier, not somebody that backed down easily if they didn't agree with something, I always stood my ground until I had depression, anxiety and paranoia, but even with those, I did have a push limit and they went over it many of times.

It was very hard to forget about them and their actions towards me. I told my aid again that I couldn't cope and things got worse again. The sleeping tablets were not working as well, because I'd be disturbed by them congregating late at night, some-nights gone 2am, it was most unacceptable or tolerable really.

I wasn't able to live in these conditions, why should a person like me have to put up with all of their trouble, in fact why should anyone put up with these living conditions? I didn't appreciate it or need it. I was suffering deeply from the depression and when it got too much it got too much. I didn't have the curtains open and it felt I was more isolated from the world, friends and family more than ever. You need quiet surroundings sometimes, I didn't have this, and it didn't make things easier for sure.

Not being able to cope with staying at home or feel able to go back to school was very difficult and challenging for me, I wasn't doing the best thing just staying at home, usually home is where the heart is but my heart was elsewhere and my head certainly wasn't in it. I couldn't put up with anymore and it was making me go through more pain and hatred for life.

The pain felt worse than my heart being torn out, it felt like I was slowly losing my own future and what I needed most, love! I rarely showed my emotion of love as it was more hate, I could tolerate people but I couldn't tolerate them or that horrid area. I was so cold in emotion, I just turned myself away from wanting to feel emotion and since have been struggling with them.

It was never my intention to leave secondary school, my qualifications were important to me as was my livelihood. I wanted my life to go how I had planned it since I was a little child, I knew what I wanted to and I didn't think of problems that could appear from life or mistakes. You never do consider problems that can arise, unless doing a PDP (personal development plan) you may consider them then, but even that you couldn't predict you'd get depression. We all have dreams about what we want to become and what we want to do in the future and my imagination went wild, I always wanted to either become a lawyer or join a television soap, it would be great to be heard and recognized. Just sitting, working from home, I remember telling myself, I am better than this I kept thinking but at other times I thought that I deserved to be bullied, but nobody deserves to be bullied and remember that if you have or are being bullied, stand up against it, get help and tell somebody close to you.

Anyone can say that being bullied is easy to deal with, but it isn't. It is similar to depression, in the sense that you have to fight it to win, I don't mean throw punches, unless you can punch, but I mean stand up against it. It can take time, but when you succeed, the feeling is well worth it, but I didn't have any fight left in me to fight the battle as I didn't think that the youths were worth the effort! I just didn't bother listening to them, their actions were dafter that anything I'd ever seen.

They could ruin the area, ruin everything that made the area but they wouldn't ruin me.

I was already a different person than what I was years beforehand but I wasn't completely different and the old Matthew had will power and courage, I still had that even when I was depressed, some days this showed through me more than others, but I did have this behaviour. I always, as I told you, would support most people with most things they did and this time I had to support myself and try to make my own decisions, what was best for me for a change. I thought running away was the best and safest option, I thought about this like I had thought about ending my life before, but it didn't seem like the right thing to do but I couldn't cope with living in hell either. So what was the best option, I couldn't work it out. Many difficulties arose from living here.

As time went past, my life got harder, it was hard for me to just sit and watch my bedroom walls that I had been looking at for months just staying the same, there wasn't any change for me in my life except maybe headaches from the loud noise, more hate and suffering. I only had that to look forward too, day in, day out; it wasn't the same lifestyle as what I had expected my teenage years to be like. I shouldn't have been shut away in my room like I decided to be, I should have been out having adventures, seeing the remainder of my friends that I had and enjoying their company was what I should have done, then maybe I would have them with me now, instead I just left them all and didn't bother with them, I guess I lost and that is what hurt. I failed my friends and failed myself!

How could my life ever get any better when I felt like this and lived in negativity constantly? I wasn't the best role model and my cousins must have needed me to be that for them, they looked up at me. If Marcus saw me like this on the holiday, he would have been upset, I'm not sure how he would have felt then, because I didn't really see him that much. I wasn't there for them when they needed my support and guidance. I wasn't there in their hours of needs and I felt really bad but I had my own hours of need and I needed mine sorting before I could even consider helping them. It may sound selfish but it's true. Change starts at home, change yourself before you try to change somebody else and my change wasn't as easy as that, I couldn't just say one day that I will be my old-self and the next day I would be a wild person. Careless of all of these findings, it takes more than socializing to beat depression, you need to want to beat it and have the support, guidance and understanding to at least try, it may not be gone forever either, I don't truly believe it leaves us, but you can try to make it more easier to live with.

I really didn't enjoy living in such a destroyed, morbid area. It wasn't always a bad area though; it was only when youths started hanging out in the area that it went from good to bad then eventually to hell! My family thought the house was great when they first viewed it; I remember having plans on how to have my room and how I was going to design it. I had known many of the neighbours for some time and they seemed friendly people, they seemed friendly when we moved in and viewed the house and then it went from this pleasant atmosphere to a violent atmosphere. It couldn't have happened at a worse time either, how could a house that I liked, when we moved in, even though it wasn't our flat, turn into a house which I didn't want to reside in or even want to see again? It just wasn't acceptable or fair and as we were new in the area it felt like a big mistake! Life isn't always fair either, those who say it isn't fair have a good point, but we can make the world a better place with our actions, our actions become our characters and we set with example after all.

The lack of sleep I had was also down to their loud noise which was contributed due to their street parties and drunken brawls, not just my worries. It was bad enough not being able to sleep but when I took the sleeping tablets they helped me get a few hours' sleep which I couldn't always have due to be awoken! I just wanted to sleep, peacefully for one night! I wasn't happy with being awoken or their noise, nor was any of my neighbours. Is it really that much to want to live in a peaceful environment? The neighbours must have been getting annoyed with the problems as well as I was, my family didn't know what to make of the sudden change but I knew what I wanted, to move out! I kept telling them this, but moving out takes time. It isn't a case of just jumping in the car and taking the house with you, like you sometimes see on T.V. I wish it were that easy though.

The area had been cursed, in my view, and it was going downhill, very fast. I remember once waking up to see more graffiti, vomit and blood stains all around the footpath visible to the window, this was on a day I opened the curtains! And this view from your only source to the outside world was awful, I thought that this was what everything would be like, disturbed and destroyed.

I felt physically sick and that wasn't the worst of their mess; they had done more than leave their blood, eggs had been thrown around the windows and one had even smashed a window. The area certainly wasn't the same and some people just allowed this behavior to continue and didn't bother about it. The problems were always happening outside my house and it seemed as if they had it in for me. I once went out for a walk as I couldn't cope with the noise that they were making and one of them again told me that they was going to kill me! I only just wanted to go out for a walk, let alone get death threats. Obviously getting these threats wasn't easy but I just thought that it was a joke and I went back into my dark room, curtains shut, into the darkness, it was nicer like that at the time. I couldn't believe that I was threatened because I left my own house, I felt panicked. It always happened to me, why was they bullying me and giving me these threats, it wasn't an enjoyable experience and I hated myself for having to put up with their remarks. As well as concerning myself with this, I had my other worries going on in my head, like what the school said about me becoming a failure and the lack of support, other than my counsellor in place. The school didn't understand and some people didn't either, it was just going around my head like all the time. I only wish I had someone I could talk to all the time, who understood, my counsellor wasn't on hand 24/7 though and my parents listened, but couldn't help in the way she, Dr. Pederson could.

The bullying wasn't a good experience or something that I would wish upon anyone but the bullies do it for a reason, the reason lies deep within them and it could be down to trust issues, problems at home or they may even be being bullied at home or by another person. It doesn't give them the right however to start bullying another person and I was a very scared person at the time they were bullying me, I just wish I had been braver to stand up to them as bullies cannot stand meeting their match, but then it wouldn't have been worth a confrontation, I just have got stronger by that experience. They didn't destroy me! They just made me understand a real-life experience, that the world isn't always a nice place.

It wasn't long until the police became familiar with the area; they were out every day and often called out in the night because of the disturbances. I could hear the police sirens before I saw the police then I saw the youths running around the area and hiding, it wasn't right and the police should have stopped this on the first occasion, their sirens gave it away that they were on route unfortunately. In actual fact this problem went on for 2 years, we were fortunately only in that area for 3 years! The problems within the area got bigger and more severe, more violence, more threats and more abuse, it wasn't an area that was habitual.

The police must have wasted a lot of time in the area as they seemed constantly out in the area but no one was ever arrested properly. I don't know why, perhaps only warnings were given? But they didn't work for sure, soft punishment if you ask me! Destroying lives through psychological effects is hard to live with. Bullying counts towards that. Whatever happened to neighbourhood trust or community? It certainly wasn't in that area!

Eventually after all of the problems, sleepless nights, feeling down constantly and feeling like I lived in hell. The neighbourhood had a dispersal order placed on it, which allowed the police to remove people from the area if they didn't live in the area or was causing problems. Many of them received Anti-Social Behaviour Disorders we were told but that didn't stop some people from letting them into houses to hide them. The landlords of the houses they went into didn't do anything either, they just kept allowing all of these problems to occur as they couldn't be bothered to inspect or sort them out. I am just glad that not all housing companies are like that and cannot be bothered, not just for the problem I had to face but for others. Otherwise if most areas were like that and had neighbours from hell, imagine what it would be like, it will be like living in hell for ever. I won't tell you what company it is, but if you've ever heard of the area down Bedford Street, you will know where I mean.

Luckily for the good citizens in the area and for me, the police did eventually crack onto what was going on and told those who rented properties and allowed them in, not too. Some still did though as it became apparent, they had been threatened to let them in, it was like that person betrayed the community and she surely didn't deserve to live there, one of the reasons why the police couldn't completely eliminate them.

I was told by my counsellor, that I had to try and go out as staying in was driving me in the wrong direction and that it wasn't helping my health. I didn't want to go out but I tried going out. I saw the youths that were causing all of the loud noise and trouble; I carried on walking past them despite all their abusive and threatening language. As soon as I was out of the area, I felt different. I felt as if something different was happening and that I could enjoy walking and visiting places more, that was until I saw a student from my school; they were hurling abusive names across the road at me. I felt intimidated and threatened. People were watching in the town as I was called a "physco and head-case" This was the first time I had been out alone by myself, for the first time in months and I didn't want to go out anymore, it didn't help me. I ran back home and felt terror. I had been publicly humiliated and people, whom I didn't even know, knew my problems. This was an awful feeling and I just remember shutting my door on the world and closing the curtains yet again. Sitting in the dark was the best thing for me, like I told you. I just shut everything out and didn't go into town again, not alone anyway. I knew my gut instincts were right, but when I was told that staying in was the worst thing to do, I wanted to change, but didn't find the whole get out of your negative thoughts easy.

The following week, I went to counselling again. I told my counsellor that I had been publicly embarrassed and that I felt even lower than what I previously had. She said that the only reason people pick on you is because they are jealous or have been misunderstood. Jealously is one of the biggest hate feelings that you can feel or experience. I didn't care if people felt jealous of me because I didn't care about them and why would they be jealous of me, I didn't want depression after all. I just wanted to live my life in peace and ignore their thoughts and abuse, but it isn't always that easy. I explained that I wasn't worried about them but their actions did hurt. I just wanted to be happy and some people in this world try their best to stop you being happy, it really is sad. Don't let them, never ever let them win or control your life! If you feel you want to do something, do it. You never know, you may enjoy it. After allowing her help, we did another plan and this time I had to try and blank people that I knew wouldn't help my positivity but increase negativity, as well as this group of people, I had to blank those who I didn't like and ignore their harsh and self-centered comments. You have to be strong and have that brave face on to put other people's opinions and views aside or at least the hardness inside of you to do so. I had previously done this method without knowing, by ignoring family members and ignoring friends and people. If I could do that without knowing then surely ignoring irrelevant people could be easier than I thought?

After we discussed this method, I was informed that my counsellor had a holiday booked for a month, so I had to cope without her. It wasn't like me to worry over not having an appointment the following week, but I was frightened that I would not be able to cope without her help for a month. Despite having family, it was her that I could talk things through with and it seemed like she understood and she came up with the answers and solutions to my depression and the factors involved in it. It was painful for me not being able to talk to her for a month but I thought I would at least try her solutions whilst she was away… they may help.

The following week was hard and it would be as I didn't have her to talk too. I had an argument about my health and refused to attend an interview to get me back into school. I was looking at the school work at home, the work just piled up but that didn't matter and I didn't want to return to a school that knew about my depression and it was clear to me that the whole school knew, well most of my year group anyway, which was around 150 students. I didn't want my life to be taken over by the bullies in that school and the bullies on the street of my home, the bullies that didn't have anything else to do but take the biscuit out of me. My actions in that school after being provoked wasn't great either, so I didn't want to return.

I had enough of people bullying me; I wasn't the same as they were all mouth and no action. I always had action but that had disappeared and I couldn't cope with being attacked, verbally-abused, hassled and victimized by them. I was thinking more clearly after the school called and I remembered that my counsellor said ignore the irrelevant people and I did ignore them. I already tried committing suicide twice and I wasn't going to try again because of them, I kept thinking that over and over and it was worth it that thought stopped me from wanting to try to do it again.

The loud noise continued and despite the dispersal order, in the area, it wasn't getting any better in the area, but the police did remove them from the area once or twice a week, I remember seeing. I allowed myself to open my curtains and I wasn't completely in the dark. I could see the daylight and hear the birds again, their singing and chirping was delightful, often not appreciated enough; it seemed good to hear them. I felt I was back in the game to feel better about myself and at least, to try and build my confidence and self-esteem back up. It was good to see the daylight shine through my window and I guess it helped show light.

I had my first home visit from the school counsellor, Adrienne, who tried to encourage me to return to school, apparently my 'friends' had been missing me. I knew that this wasn't true as I didn't have any friends and I wasn't going to call anyone 'friend' without thinking again. I didn't want to return to the school and I expressed concern, the school counsellor thought I was being unintelligent as I wouldn't be learning anything new and she urged me to return or action would be taken. The threats that they try and use to get me back into school, it isn't fair. I didn't want my parents to be in trouble with the legal system so I returned…but that wasn't the correct thing to do and it was one of the hugest mistakes I ever did make.

Once I returned to school, the entrance felt like the entrance to hell. I could remember my first day and all the painful hurt I felt over leaving my other school and piers behind. I remember the tutor who I didn't have much like for but could tolerate waiting for me and she was belittling. I couldn't bite my tongue with her and spoke my opinion of her; I got in serious trouble with my head of year because I spoke my mind. She was the one that started this feud by her mockery of my condition. I wasn't going to let her and thought I was being a victim of more harassment. Teachers are supposed to care and understand, not try and get gossip about the students for their tea chats in the staffroom. I didn't care who she was, teacher or not, nobody was going to do that to me, she was so stuck up she thought the sun shined out of her but this wasn't the case. She was just a little lady, who eventually got misfortune. I guess she shouldn't have been so judgmental of others? It isn't wrong feeling like that, when you have felt like I did back then. I think deep down she had some compassion but it was hidden as she was scared of something, she always seemed to keep her private affairs to herself, unless it was about her daughter and then she would tell you her life story and how great she is or what she was doing.

I had a chat with the head of year.

My thoughts about not returning to school proved to be true. The first hour back and I was in trouble. The school didn't care about my depressive state and put me in isolation all day! I had to be separated from everyone in the school and shut in a room with a desk, I'd have been better off at my house. They had no idea whatsoever! As long as I was back in school they were happy. I didn't want to be there and I refused to do the work. I don't know why I returned but I guess it was to try and I couldn't think to see my parents in jail for this. I did go back, but when your tutor does that to you it is not fair or acceptable! I told my head of year that I didn't want to go back to school but I was forced to return, I then told them about what the tutor had said and they seemed to agree with her to an extent, this wasn't fair. The teachers all stand for each other and the students are ignored. Again, not enough understanding of mental health, vulnerability is a part of it and being insensitive or degrading, doesn't have positive effect on us who have it. I wasn't going to let a head of year tell me what to do and I walked out. There is a point in life where people are wrong and if they cannot admit it, it makes them a bigger coward than ever. This tutor had been at the school longer than me but it doesn't give her any given right to judge my life and my hatred for her just grew…

Once I walked out of their office, I went to find the school counsellor to let her know what had happened. I saw many shocked students looking at me and their heads turned as they started chatting… I just kept walking and I bumped into the group who I called friends. They were also shocked to see me and questioned me as to what I was doing back and if I was ok? I didn't want to make a fuss and told them I was fine. I had to sort this problem out alone, so I carried on walking past them; I didn't stop and didn't chat to them for long as I was so determined to prove my point. I went to the reception desk and asked for them to contact the school counsellor. She came down and we had another meeting to see how things had gone on my first day back; I told her instantly what had happened and she wasn't impressed! It appeared someone was actually willing to help me in that school, much to my shock I was being helped and she could understand why I couldn't be at that school, finally I was heard!

I had to give a statement about what had happened and what the tutor actually said. I remember writing everything down, I even wrote down what I had said and I didn't regret saying the things I said because I was provoked and had taken enough grief from people throughout my time at school and home-life. I also informed her that if I was expected to return to school then why was I being spoken to like this. I also asked her a question which she had no answer for:

"How do you expect me to return, when everyone knows about my state?" I liked keeping my life private and it was exposed.

It was like I had asked her a rhetorical question and no answer was needed. But I wanted an answer, so I repeated my question and a humble "urrrm" was given.

After our chat I was told to go home- which I felt was ridiculous especially as the school wanted me to return. What about the legal action threatened to my parents if I didn't return? I wasn't going to argue as I didn't feel stable enough and I didn't feel happy at the time to stay. I kept a blog of the bad things that were happening to me and I never put any good things in the blog. It was a pain for me because I knew that I couldn't contact my counsellor who I trusted and needed for help as she was away.

On my way home I saw a friend who I hadn't seen in ages. I ran up to them and felt happy to see them. I hadn't seen much of my close friends as I stayed at home and didn't think I had many left if I am honest with you. I was overjoyed to see them and they started talking about how their life had changed for the better, they have joined in different clubs and been enjoying their life. They asked me how I was and what I had been up to…. I froze and stood still. I wasn't going to lie to them about what had happened. I told them everything that had been going on:

"I have been receiving professional help as I cannot cope alone. I almost took my own life and left my family in pain."

I hated those memories then and I do now. They made my mood worsen so after our brief chat, I ran off past them, got home and slammed my front door shut….another dark moment which I couldn't control.

When things got bad they really did get bad. No matter how low in mood I was, I actually wanted to be happy! I tried to cheer myself up when I got home by watching Eastenders which usually cheered me up for a while anyway, as I was watching other people live life and it made me forget all about my worries and my own life but I couldn't be happy as I was annoyed with the school for treating me like an invalid person and causing me further distress, something I needed to try and cut out of my life. I was ever so fuming with my tutor's sarcastic and blunt comments, they were uncalled for and I believe that the only reason she said them was to make me feel low. She was the sort of person that didn't enjoy other people being happy, she could be the only one that was happy.

I just kept bottling my thoughts and my anger and it got worse and the feeling of hate built up more inside of me. I wasn't enjoying watching the television anymore which wasn't like me. I had stopped eating as I was paranoid about my weight; I still wasn't talking to people as openly as you would want somebody to do and just stayed in my room once more. My counsellor still was on holiday, so I was unable to talk to her about my frustrations. I spoke to my mother though about how I was feeling, she could understand and tried her best to help me. She was also fuming with that school and regrets ever letting me attend that school, it wasn't her fault though, despite their Ofsted rating, when an individual case like mine is hidden, any rating can be given. That's what is so hard about making the right decision, you may know the building on the outside, but it can be full of woodworm on the inside. It is hard to try and live someone else's life and the feeling was like that. My pain was hurting me so much I could barely talk about what I enjoyed anymore and the motivation side of me was destroyed, I had no strength left to fight.

I did have so many hobbies and enjoyed so many things which I used to enjoy before the depression took over me. I wasn't the same person that my friends or family knew I was like a person who was always moaning about everything and it was beyond control. What was the worrying thing though is that I didn't even know myself!

I feared everything and everyone. My brave face had gone for good and I wasn't confident at all. I didn't speak to my parents about everything that I was going through or feeling and I felt very isolated from everything. As I wasn't eating I got very bad headaches and felt I caught colds easier than previous. Within a few weeks, it must have been six weeks, I lost 2 stone. The colour of my skin was much whiter than it ever had been, I lost the feeling in my hands and I had trouble sitting still, I was very fidgety. I was still hearing the strange voices in my head and these confused me, I saw shadows around me and felt cold presences. I didn't want this to continue, so I called my doctor and asked if there was anything else that he could do for me. I then had to go to see the doctor. The doctor then increased my sleeping tablets and told me to take 1 more to the dosage I already took before I go to sleep as I couldn't sleep. I was on anti-depressants and sleeping tablets- both of these dosages were high and I didn't feel comfortable taking tablets. But they needed to be taken, otherwise I may have felt a lot worse.

The tablet's dosage increase wouldn't block out all of my worries and the loud noise caused by the youths though surely?

After a night or two of taking these tablets, they kicked in and I started sleeping for a few hours more than I had done before and it did help, I wasn't so tired and grouchy as the increase seemed to help me. The dosage increase allowed me to have the best night sleep. When my head hit the pillow I was gone, my eyes finally shut and I could rest. In the sleep I had this dream; I saw a good memory of a time I shared with a deceased friend. I had a great time and enjoyed lots of memorable moments with her. She was a very close friend to my parents and I… she would help whenever she could and where she could. She died of Dementia and this left me devastated and thinking about death. The death worry I guess was caused by losing someone close to me. I cannot cope losing someone without saying goodbye to them or thanking them for everything they did for me, it was shocking, life is cruel in that respect. So always make the most of people around you, before it is too late. Despite the worrying, I felt happy to be able to see the memories that we shared and the great times we had. It was a very unusual signal for me and I thought it was a sign trying to help me…

She must have seen me struggling and wanted to help me in memory, Muriel was always a person that put others before herself and without this memory/dream, things could have got worse, the doctor helped me by increasing the sleeping tablet dosage and the memories came to life in my dream. I know sometimes dreams seem so real and this dream felt so real that I could feel her presence. She was the voice talking to me in my mind. Maybe this dream was an encouraging one, all dreams have different meanings after-all.

I discussed the dream with my family members, most of them thought I wasn't stable at the time so they just thought that I hadn't had the sign but I knew I had….the voice in my head spoke again and it was as if the voice wanted me to change. My door slammed shut and the hall mirror and photograph of me smashed on the floor; I was glad it wasn't me going to be receiving 7 years of bad luck, this was very surreal.

It was definitely a big sign for me, I saw Muriel, who was around for much of my childhood for my family and I, she wanted to help and support me in my dream and I heard voices and with my picture smashing to the floor with a mirror, this was a sign which told me to look at my life closely and change. I thought the sign was an instruction from my friend from up above, these things certainly do not happen every day and it was something that made me try even harder to beat the state I was in, the fears I had and my depression and gave me some encouragement as I thought she had gone forever, even if I couldn't see her properly I knew it was her, she felt so real. It's always when you try to figure it out, you awake and the dream ends that always leaves you thinking- did that just happen?

I didn't care if other people couldn't or wouldn't believe me, I knew this feeling was real so I got out of my room and I tried instructing the change I wanted to see and went through everything I had, it was time to sort out my life, starting with me. I was feeling strong and more encouraged to beat the depression, it wasn't done as easily as this though and I was on the anti-depressants but I wouldn't be beaten, the pills helped me be a little more positive about life, but they couldn't control life itself or every single factor. Nobody was going to make me feel low again and I was sure I was changing.

After making some changes to my life, I believed I was getting happier so I decided to take myself of my anti-depressants without consulting my counsellor or parents, she was away after all and I couldn't speak to her so it didn't matter. The doctor didn't put me on the tablets so I didn't tell him either. After a few days without the tablets though, my mood went odd and I remember having the same moment as before, where I couldn't sit still, I had to go around my home throwing away all of the memories that I had and most of my photographs were cut up because I couldn't stand seeing myself in them. I had changed and didn't want the memory of how I used to look. I then moved on to burning the photos and my possessions. I wanted a fresh change and that meant a new me, I couldn't change how I felt inside but tried to change my fears by making the outside look different. Changing who I was though didn't make me happy at all and I was again sinking in mood and not getting any better. The change in the photo's is what got to me the most I believe though, I was happy back as a child, full of love and beaming, I wasn't full of sadness, have puppy-fat or acne. It really does affect people.

As I wasn't feeling happier I questioned my family about their loyalty to me. Nothing was left unsaid as I confronted them all about some of their selfish ways; they might not be able to see my depression but I could, I had to live with it for many years after all and at this stage, I had had enough of most things. Their selfishness was mean and felt like the ultimate betrayal, however despite them not all helping me; I was out to prove that I would not end up having depression for all my life! I did love my family despite their opinions; their opinions were just worthless to me on this occasion.

I was trying to be a happier and changed person but it is hard to change and even harder to change when you have depression. The happiness was never there, it wasn't within me for some time and I couldn't feel happy about anything I changed. I changed my hairstyle, brought new clothes and tried being a happier person than I actually was; I had an ego that hid my fear and hatred. The ego was bigger than my personality and me.

Chapter 11, Life Continued to Change

I did over try too quickly to change as many things about me that I could and I do not suggest doing this. Fortunately after this downfall, I received a letter from my counsellor announcing that she had returned from her holiday and she booked me in for some more sessions with her, I refused to go at first though. I thought that I had been coping without her for so many weeks and it wasn't going to change because she returned from her holiday. I wanted to fight the depression alone without friends or family.

My depression was like a ticking bomb, I had so many negative thoughts and regretful moments and some I couldn't even see or understand. I didn't accept that I was getting worse; I convinced myself I was getting better and enjoying my life. I had always tried to convince myself I was fine, even when I wasn't. I guess it is because I don't like admitting I need help. Why I thought that I will never know as I wasn't happy, I didn't see any friends to meet up or spend time with, everyone else was always meeting up and where was I?

As I wasn't on my medication I wasn't really sleeping again or getting any help with the depression. I again turned to my old enemy- suicide!

After two unsuccessful attempts, I attempted to jump out of my bedroom window. I opened my window and was about to jump out when my mother walked into my room and demanded I stopped. She knew I wasn't in a very good place and everyone in the area thought of me as a mad person, it wasn't a good reputation and if my mother didn't appear then I probably would have jumped 20 feet, from my bedroom window. She was one of the people that wanted to help me, she had seen me before when I tried and it really wasn't fair on her to see this especially again. I wasn't being selfish, it's the mind thoughts.

She contacted Dr. Pederson, who advised her she had sent an appointment out and I again had to go to counselling sessions. I don't think my counsellor was happy with my progression as she booked more sessions in with me, her diary had me in every week and even if I couldn't see it, I must have been in a really bad state, fortunately though, she was supporting and had the time I needed so badly.

The bad state was seen by everyone; I had no confidence and always stared down at my feet attending counselling and this was just a short walk from the car-park across a road. I was nervous that everyone was watching my every move and I didn't feel comfortable looking at their faces. I was scared and didn't want to communicate with them or acknowledge their presence. In the counselling clinic it was worse as my mother always booked me into the receptionist and she must have thought that I was a shy person and I hadn't always been a shy person. The depression takes over and sometimes destroys who a person is; all of my skills had been taken away as I was too afraid to use them. My confidence turned into shyness and my ability to talk openly was also banished.

My counsellor came down immediately after the receptionist contacted her. She greeted both my mum and I, and welcomed us to her consultant room, mother came in this time, and she hadn't been in for all of the sessions. I could tell she wasn't happy with my progress or my failure to attend her previous booked session. I did go through all of this previously as discussed, but I couldn't change my fear of people and I wanted the help to do this. She had many cases of people with depression as I remember asking if I was the only person to have depression as a teenager, many people with depression aren't actually diagnosed as most see them as the 'teenage mood swings' but some moods are much worse than that.

Again we discussed what problems lied within me, it was easy telling her what my problems were but harder to sort them fully. My main problems were:

1. Body image; I hated my body deeply, I didn't enjoy looking in the mirror so I didn't, I couldn't accept the change in my body, I had acne and it wasn't a pretty sight and I didn't like putting on weight so I didn't eat, which resulted in the 2 stone weight-loss.
2. Lack of social life; I didn't see any of my friends on a regular occasion, I was like a prisoner in my own

home as I didn't go out and was constantly in my room.

3. Fear; I had a lot of fear and hate for the area I lived in, I didn't appreciate being a laughing stock within my home area, I hated it so I refused to go out and this along with the other problems made my life harder than it needed to be.
4. Paranoia; I was paranoid that everyone was talking about me, I was scared and worried that I was always being watched and this worry never left me and it was always with me so I couldn't trust anyone or look at anyone. I didn't speak to anyone I didn't know or anyone I did know.

After discussing my main problems, my counsellor saw it was more than a plan I needed it was more therapy with another counsellor, a counsellor who worked with depression suffers for a while and appreciated all what happens and comes up with solutions on how to sort fears and become more able to live with depression person. When this was suggested to me, I initially refused to change counsellor but it wasn't my choice it was essential I changed if I ever wanted to be more able and more capable. So I eventually agreed.

I was always worried about not being able to be the same person I once was, what happened if my strengths were my weaknesses? What happened if I carry on not enjoying being me? Or even my life? I couldn't answer these questions as I couldn't see into my future and even a crystal ball wouldn't! After agreeing to see the new counsellor, I was told that she would be in touch and that I had to take my anti-depressants even if I didn't want too as they are there to help me get better and not worse. Mother must have told her, I am glad she did, because she really knew as she controlled my medication after my suicide attempts.

So after that, we left the clinic and returned home. It was awful being at home as I felt locked in. I wasn't locked in but that is what it felt like. I hated seeing the troublemakers when we returned home, they constantly were in the area, even with the dispersal being served… They destroyed the neighbourhood and everyone wanted to move. My family finally decided that enough was enough and the neighbours and they would come up with a plan; they would call the police and destroy the parties they were having on the street. My neighbour always had loud parties and this was what kept me and others awake also at night, when I could sleep it wasn't t tolerable, it wasn't so bad when I couldn't sleep as I was awake. The youth's bash' always started in her home; her window was smashed twice whilst they partied, again that window shattered. It wasn't right but the police knew exactly what was happening and it didn't take long for them to attend the scene of hell and remove the youths. The woman was given a further warning but never arrested which shows how soft the police actually were with her. The police left the premises moments later without making any arrests and the neighbours were furious. They decided that they would throw their own party on the street to annoy her however it backfired on them as she joined in and got very drunk and even attacked a neighbour. They had a street brawl and both bruised each other, it was getting worse and again I thought of suicide, the negative thoughts kept appearing again but somehow I could ignore it, more than last time. I saw a flash memory of what my family would see if I did attempt it and thought again I don't want to leave this life, before it gets better again.

I just opened my window and had a go at the silly self-loathing woman; she had annoyed me so much that I had to unscrew the bottle that I had been getting full of annoyance for her, life and other problems I faced. I went ballistic at her and made sure she knew what I thought of her, she was too drunk to even stand up but I thought I would add to her problems like she was doing to mine. She might not have even remembered what I had said but I felt a lot better getting that of my chest. Usually I could keep myself from getting agitated in public, but again that skill had gone and I was determined to get them back as I wasn't going to let those skills be taken from me… for good.

The police then turned up again, still doing nothing about this idiotic fool and just escorted her into her property to avoid public nuisance. She should have been arrested for drunk and disorderly behavior but she wasn't and that was somewhat annoying. If it was my age generation of 20, we probably would have been!

I had taken my sleeping tablets contrary to this altercation and then was able to sleep for a few hours before having to wake up for my meeting with the new counsellor, which had been scheduled very quickly. I was feeling worried about meeting her as I wasn't sure what she was going to be like, or whether she was going to judge me like most people judged me, just because I had this problem it didn't make me inhumane. I thought long and hard what I was going to tell her and couldn't get back to sleep even a few hours' sleep was better than none at all, I lied awake thinking and thinking, the hours went slow and my nerves started playing up.

The time eventually went round to my appointment time, I was waiting in the reception area with my mum and the new counsellor along with my old one came down to introduce her and I felt very awkward meeting a new person as I didn't want more people knowing about my business, she however looked very trustworthy and I introduced myself. I was sure she knew about my problems but she asked me:

"What has been going on and why do you feel the way you do?"

I didn't want to tell her the honest truth but thought I better had as she had my notes and if I wanted the problem solving or at least repaired somewhat, then I had to be honest with her. I told her everything that had been going on since I was referred to the clinic and all the events that I went through such as suicide and hate for family and jealousy of friends. I then told her about the area I used to love residing in turning into hell and the thought of living made it hard. I could see by the way I replied that she wanted to do something immediately. Although I didn't know her, I thought she would do everything she could to help me.

She reviewed what had happened previously between me and my other counsellor and she suggested a few new things like introducing a bedtime. I thought this was very ridiculous as I wasn't a child. I asked her how I could have a set bedtime if I couldn't sleep. She was honest and said that routine increases the chance of success and that if I was going to be successful in dealing with each problem I needed to address them individually and adapt to fit around them. I thought the idea was still bonkers but agreed to have a set bedtime and take the sleeping tablets 2 hours before I went to sleep.

The second thing she did was adjust the planning we had already made and give me a new plan and this plan was the following:

- **Stay strong and be positive- try and see the good in things even if they are bad;**
- **Ignore any bad comments and think positive;**
- **Don't allow the bad things to bring you down and remember that you can be better than the bad;**
- **Enjoy your family and try to mend things with them;**

She then acknowledged that some of the plan may be hard to follow as it is increasingly hard to see the good in anything when you are used to seeing the bad in everything. This was my first task set by her- try and see the good in things.

The plan seemed easy enough to follow, but it was hard to put into place. I tried following the plan and tried seeing the good in my neighbours, some had good qualities about them and some didn't, but it wasn't up to me to judge them, I just had to find good qualities about them and see the good.

I took notes down about what good I saw in them and how I could see it, when I set my mind to it, I could actually see the good and see their happiness despite the recent events with the fool in the neighbourhood. Most of them were going about their own business and looked positive, more than I did anyway. I left the house to have a chat with a neighbour and remember one of the youths calling me "nutter".

I just smiled and replied "yep that's me." I just ignored a rude, worthless comment. I was able to control my negativity by mocking their ridiculous bitterness. It was very different being able to laugh of their comments, instead of letting it bother me. I kept re-reading the plan to make sure I was following it to the best I could, I saw the part about ignoring bad comments and staying positive and if that wasn't what I did, then I don't know what was. I felt positive about being able to do this and I made sure I kept this up, ignoring the comments and just letting them fly in one ear and out of the other was something I struggled with previously, I could ignore the person saying the comments but couldn't ignore the comments as I would bottle them up, the words did hurt sometimes.

I still couldn't trust half of the people in my life with my problem and the depression hadn't gone away, it was more controllable with the help of the tablets and professional help which I was receiving. I was able to discuss more of my problems with certain people, whereas some wouldn't listen and some didn't have the time or day to listen to my problems, instead their problems were worse, it wasn't nice dealing with them people as they wasn't bothered if I got better or worse, they seemed like they only cared about themselves and I still feel a little resentment towards them now.

I thought I could trust all of my family members and I expected that they would be there for me in my darkest moments but they were not and I needed them even if they didn't need me. I tried telling them about my sessions and what I was trying to achieve from each session and they looked but didn't really comment, why would they be like that? Before the negatives came into my mind though, I put on the positive thinking and just thought they are not worth it if they are going to treat me like that, I ignored them like they did to me and despite that being a negative thing it actually had a positive outcome as I was disinterested in their views and wasn't worrying myself over why they were being like that.

My neighbour was still annoying somewhat but the negative comments about the situation were all wrote down instead of thought of constantly and I just saw her as a big failure and that encouraged me not to end up like that or follow her paths. I wanted to make a success of my life and dealing with the depression was the biggest success I could have ever hoped for in my early stage. The neighbourhood was never the same after that argument or youth congregation and many didn't want her to be in the neighbourhood either but she wasn't going to move and she wouldn't be happy if she moved so she stayed and decided to cause everyone else hell. They didn't want her as a neighbour and I certainly didn't as her house was semi-detached onto mine; the noise from her was like living next to a field full of cows which woke you up every-time you got to sleep, she didn't deserve a house here!

I was determined to not let her problems contribute towards mine but I was allowing the stress and idiotic actions of her to bother me, it wasn't fair as it felt like I was living a life that I didn't want and I just wanted my old life back without her presence in it. Our other home before we moved into that area was lovely, it was nice and spacious and the neighbours had boundaries and respected each other, we all used to meet up and arrange events such as sport events and get together. I had everything I needed there; I was close to my schools, friends and the town. I didn't get bullied as I didn't suffer from depression at the time and I wasn't patronized by lazy scalawags as there wasn't any criminal behaviour near my home. I had dogs which I enjoyed walking and I was never in my room all the time with my curtains closed either ,as I could bond well with my family and discuss anything. Things changed though when we moved and although we cannot live in the past forever, it would have been an easy option and changing time could change and make me live back there, as well as what I did and what I wanted, but as we grow we get achievements, regrets and wish we could change things that we cannot, sometimes it is unbearable to think about the past as the past is the past and something's should remain in it. But that is life. We learn on life's journey.

I realized that the past wasn't going to come back and I had to accept that I had to move on, it hurt me and certainly wasn't worth thinking about all of the, what ifs; what if that was changed and what if that happened differently. Things happened for a reason both good and bad and I needed to accept that I had to fight the neighbourhood problem rather than hide from it. My old flat may have been better but there was no going back, so I was adamant that I would improve my situation and I decided that the best thing to do was cause problems for my neighbour like she did for me.

I was fuming with her lack of intelligence and declared her my worst enemy. She was more annoying than a fly buzzing around a room. I decided that to get my own back on her, I firstly played my music loud and I liked classical music at the time, it helped me relax, so I opened the windows and put the entertainment system on full blast. I could feel the vibration in my house so I was sure that the sound waves would be going through her house. I enjoyed blocking out her noise and listening to my own as mine was much more relaxing and more influential than listening to her raving parties. She sent a youth round to ask me to turn my music off as they didn't like it, I thought no why should I, so I told them to get out of my garden and get stuffed. I wasn't going to let her tell me what to do especially as she didn't listen to any of mine or my neighbour's comments. She was really ridiculous thinking that she could give me instructions and demands.

She must have got annoyed with me, but I wasn't about to let her take the neighbourhood over and destroy it as it was nice until she allowed the youths into her property despite breaking the dispersal order.

I got annoyed further as the school contacted me again and requested that I attend some pre- GCSE mock exams. I wasn't happy staying at home, so I agreed to attend them despite the problem I had with my form tutor, she didn't have a clue about me attending Maths. I wanted to see things more positively and I wasn't able to do this whilst being cooped up and alone at home. So with a little worry of fear of returning to school, I attended my classes and saw the other pupils looking at me, it was like I was a ghost or something, the feeling of seeing them all was surreal as I hadn't seen them for months. I was approached by some of them and remember the great feeling I had when I saw my geography friend, Brooke. I had missed some of them and it was encouraging to know that their help was being shown and offered to me. I felt like they would support me through my bad time and I did hope that they would help me despite my constant worries.

After having so much fear and despise for the school, I actually enjoyed being at school for that lesson as it got me away from the headache of a house I lived in at the time. I saw some of the teachers who had also been worried about me too and I certainly wasn't forgotten. It was like I had returned from a long holiday and everyone was overwhelmed to see me again. I had been studying the materials that the school had sent to me so I was aware of what the curriculum was, even though I had a stack of work that had to be done. It was very different compared to my first day as that transition was hard and the work was very different and confusing. I had the support that I was longing for and that support felt great, it was mainly peer support though, rather than teaching support. I wasn't punched again and I didn't actually see any nasty people in the school, it must have changed!

The mock examinations went okay, as they seemed understandable and I was glad that I felt better over returning to school for the exams. I had my exam timetable given to me and the exam subjects were okay I guess. I did enjoy being back in school and in contact with so many people that I hadn't spoken to for a while; it was brilliant to see people who thought something bad had happened to me. I remember someone thinking they were funny after they shockingly questioned me:

"OMG you are still alive. I thought you died. What happened?"

I wasn't sure what to make of this absurd comment so I looked at the person and smiled and then I replied:

"Are you talking to me? Now why on earth would I want to die when I have waited so long to see you look that shocked?"

I didn't care for a response so walked on past this person. I wasn't letting anyone try and dig my bad time up and none that tried could be able to, they didn't know the complete story anyway. I was too cold hearted to listen to them and I had a new attitude, ignore other people and their comments! I was able to ignore and criticize people back without giving a damn. I was sick and tired of being the quiet fearful person and thought that being back at school wouldn't be the same journey I had feared but a journey I wanted. I was back with the fire inside of me and the exams were about to begin.

One by one we were called into our examinations and I could hear people talking and it made me feel happy. I was happy that I was back and I certainly enjoyed being centre of their attention! It made me feel missed and that feeling was great. I found a new direction and that direction was winning. The exams went well I thought, I may have made a few mistakes in the exams but who doesn't? I wasn't going to fake my answers or rush through the exams so I took my time and recall looking up to the high ceiling in the hall and thinking my answers over before I wrote them. It is amazing what I remember.

My timetable was much shorter than most of the students as the school cut it down to meet my requirements. I wasn't able to do all the subjects at home that were needed for the exam as I needed guided hours in school to learn them. The subjects I did in GCSE were: English, Mathematics, Science, Food Tech, History and Geography. I enjoyed most of these subjects and regretted not being in all of the lessons. Some of the exam questions were solvable but felt unsolvable. Before I knew it, I had been back in school for a while, I didn't even notice and this was when I noticed that the confidence in me was getting bigger. It had made me seem a little less frightened and a stronger willed person and I was getting my studies back together or trying at least. I was enjoying a part of my teenage life that I never enjoyed previously. I was more relaxed at school than any other place I was.

When my school days finished though that is when I felt regretful that I had even attended school as I was returning to that living hell! I wasn't sure if the problems were happening in the day or just when the fools were about, but either way I didn't want to go home and I often went through a garden to get home, it was a through walk which allowed me to miss all of the bullies and ignore their presence. It was great being able to do this until the person locked the through gate so I was unable to pass without them seeing me. I was petrified about what to do next as I had to walk past them to get to school and home, I did this however and realized that it wasn't all as bad as I thought… well it seemed as if they had got better things to do or another victim.

If the police saw them, they would be sent on their way so the amount of time they spent around the area was in that woman's home. Things eventually went really bad again; the youths were drug dealing in her property, in taking large amounts of alcohol and smashing glass bottles against my home. I went back into a state and despite the police inspecting they were never able to catch these people as they just ran and hid. I thought things were getting better but this area would never improve and it didn't feel like our much loved house anymore, it hadn't for a while at this time.

I had school the following week as the exams were continuing. I guess the exams came at the right time, I was escorted to and from school and was able to feel comfortable and less freaked out by being given this option, it did help me and being at school kind of took the worrying away from me. I saw the Adrienne again who was glad that I had returned to school and she wanted to continue doing some more home visits as well. I was being listened to, I wanted to know what her plans were with me and I respected her. She like my counsellor, wanted to help and she helped me with my school timetable and she also helped me with my examinations. I was glad to have the extra help I needed as most of the teachers who knew me, never had the time or day to help or listen. I was glad that the stranger I didn't want to talk to became one of my biggest help. Never judge a book by its cover applies here.

Despite her not being a close friend or me knowing her fully, I was able to know that I could trust her. It was nice to have someone as eager and determined like her to help me. I really appreciated her help.

I was attending school, attending my exams and the tablets were helping me feel a little more positive and I was able to sleep a little better than normal. The only problems I had was not being able to see the good in people and thinking that everyone's comments were aimed nastily at me and of course the hell of my neighbourhood. These problems needed sorting out, so I decided that I would go to my next counselling session without causing a fuss. So I wrote down my problem and read the plan over and over again and put my comments down and what I was finding hard. It was easier to write down what I was thinking and sometimes I remember thinking that talking about problems made them worse as you see them come together in a picture. But it is easier to see and confront them, before they escalate and worsen. I now encourage those who ask me for my advice on my website that they should always talk to close relatives or friends, expressing their worries or concerns, to receive genuine support. A supporter for somebody who has mental health, does go a long way.

My new counsellor was glad I went to the appointment, I remember her telling me, and I told her that I had attended my exams and thought I did well, which of course I eventually found out I didn't do so well and I didn't sit all of my exams. I also told her how happy I was to see the people I had lost contact with and they were happy to see me, as I thought everyone was against me I put seeing anyone of for years! I made the right decision to get back and face them. Not all of them hated me, some really did care, they weren't all two-faced in telling others my problems, and some of them really did show compassion. I must admit that I did shed a tear telling her this and it was the first time I felt more-satisfied with life for a long while. Despite there being positives in my improvement the negatives still were there- the comments from the bullies that were aimed at me and the trust issues as well as the area I lived in at the time, confidence and trust issues and low self-esteem. I just wanted to be more able to trust people and adapt to different people's personalities, so I could at least feel involved in their lives. Of course I wasn't liked by everyone and the fact of life is that we don't get on with everyone we meet. Not everyone we meet on life's journey, we become friends with. I'm not the only person with that belief I am sure. I told her that I was trying to build my trust confidence up in people, by trying my best to communicate more at school, I did attend Year 10 of school and that was the exam preparation and exam year, as well as Year 11, which is one of the other years I wasn't able to attend secondary school, which was the important year for the big exams, which I failed because I didn't attend. Although this was unfortunate and classed as a 'failure', it actually became a turning point for me and I would not change anything about this, because at the time in my life, I couldn't see things differently, whereas I can now. I don't regret this, I just look at this moment and feel motivated as to where I have become today.

Trying again to tackle my life's problems was something worthwhile. I was more determined to try new things and trying was something I liked best, like I told you, I am a trier not a slacker. It was unfortunate for me, that I didn't feel ready to go out and socialize, because this could have perhaps helped me further with confidence and self-esteem. I still thought everyone was talking about me and after that vile comment that was shouted at me in the town, I felt worried that more and more people would find out that I suffered from depression and it appeared to me that little is understood about mental health. So even though I was trying, I was limiting myself to what I could do, without it being an intention, more of the way I thought. I thought my counsellor could help me with this issue, but all she could do was give support & guidance to me and she came up with a strategy. Dr. Pederson and I came up with many strategies and ways to cope with my depression, so having another strategy made it seem like my life was become a strategic plan! I was grateful for them, don't think I wasn't, but each one had so many different things to act on, it became surreal. This strategy was a guide which was clear. The strategy insisted that I encouraged my confidence by being more charismatic and practice speeches in front of people or a mirror. I hadn't been able to look in a mirror for long before as I didn't like what was looking back at me. I wanted to develop more confidence and boost my chance of feeling better. I listened to her advice and after discussing the continuous problems I was going through; I had to follow another strategy like I did the other one. So I did it. I looked in-front of the mirror, which I hated looking in and saw my reflection. To say I was happy with my appearance is an understatement, I hated it, but nonetheless, I told the mirror about me and felt I could talk about myself to the mirror, but not so well to other people. This was to come later on. I saw somebody I attended school with, after we left and they couldn't believe how much I have changed from then to now, I have little acne now and certainly am not the size I once was, which has made me a little more confident, although not totally, but this shows that puberty does come and it does go, so never feel like it will not, even if it does take a while to pass on.

I could stand in-front of the mirror for several hours, just talking to myself and looking at my reflection. Being able to do this was advantageous, because I had to admit who I was and that was the schoolboy who had acne and a little extra weight on him than others. But that shouldn't mean I wasn't normal, because nobody is normal and differentiation does happen, that's life too! The main problems I was facing were caused by home stress and failing to trust. I couldn't change these issues as I didn't know how to, I never had to try to trust before, I could just trust and it was that easy. Trust is delicate because once broken, it can never be fixed easily, sometimes they say 'forget never forget', it does hurt when trust is broke and we have our trust broke at least once or twice in our lives. Which is unfortunate, I acknowledge, but also it opens our eyes to things we often cannot see and helps us learn. It seemed like I was being more tested than before and I didn't like that. However, I may not have liked that at this time, but I am glad in some respects that I was tested, because I now do not give my trust out as easy as I did as a child, I think twice and don't trust people I feel are unworthy of it, or I know do not really care for me, finding the difference between 'real' and 'fake' people sure is worthwhile and this is why I am glad my trust had been tested.

It did feel as if my family still wasn't happy to be there for me and I felt like it was my fault and that I had, somehow, asked to have depression, when of course it wasn't my fault and I certainly didn't ask to have it. I couldn't understand and still I did not know why they wasn't helping or supporting me. I asked my mum why they were being like that, they all didn't have any understanding of what it felt like going through this and they compared me to other teenagers. It wasn't fair or right. Making this comparison is like comparing everyone against everyone, hoping everyone is the same, I'd hate to even think we were all the same, the world would become so boring, many personalities lost and people all the same would be tiresome. I only wish for more understanding in our world!

After speaking to my mother about why they couldn't understand me and had been judging me, I felt they were judging me and sure feel that way now. My mum answered that "it isn't your fault it is theirs as they are the ones thinking that way." She was always helping in whatever way she could, be it support, encouragement, taking me to my appointments and standing up for me. The role of a mother is one of the greatest in life and she has lived up to be a great one to me and my sister.

I can understand why they had they had their views, but them having their views is different to supporting me. I thought that they would have at least helped me see the good in life. Some of them just obviously didn't give a damn about me and that pain of insecurity and constantly being judged, never did really leave me and even now I am still slightly judgmental. I knew what family members cared and what family members didn't, let's just say that. You realize the ones that count eventually though.

I knew that I had to get through the new plan that my new counsellor, Vicky had set me, I didn't go through the plan alone though, I had my mum who was a great help to work with. She was there for me to talk to and as she always attended my sessions with me she knew what was said and what I had to try and work on to improve. She came up with a new technique and a way that could help me build more confidence. Go out and about with her. I felt more able to go out with her as I wasn't alone and wasn't without someone to talk to and be apart from, I did feel more secure going out with her than alone. Although she didn't always come into all of my counselling sessions, because Vicky requested some were just 1:1 sessions, I always told her what had happened or what we needed to act on.

The first outing we had was to town together, nobody shouted at me this time, which felt great, nobody knew me and it was weird that my fear of being out, and in-town was down to an incident that happened so many months prior to this. I didn't know if this was because my mum was with me or what. I didn't see anyone I knew so I felt differently. I didn't just not go out because of this incident though, I didn't feel able to go out and when I did this incident happened, which left me not wanting to go out. We then returned home and there was more trouble, graffiti had been sprayed all over the area again! It was on the bins, on the walls and even on some people's windows. This was a disgusting sight to see and along with the smashed window which was covered with cardboard, it made the environmental factor look like a pile of muck and this was just after the environmental people came out to remove the graffiti! What a waste of money this was for the taxpayers to endure! After seeing all of this, I was just glad to get in.

"Did that help Matt?" Mother asked me.

"I think it did, I really thought that everyone knew my every move and everything about me in this town, I guess that outing with you proved me wrong." I replied.

I felt much more positive after this outing. It was nice to be proved wrong for a change and to see that not everyone was against me, the fear of everyone knowing my every move, sure also was just that, a fear and not reality. Paranoia can sometimes take over, when we ever don't know it, I do admit it isn't easy to have paranoia and this is why the need of support by others is paramount. After this positive outgoing with mum, the next one we had was a visit was to see my Great Nan, Doreen; I used to enjoy spending time with my Great Nan as her company was wonderful and well appreciated. She was always helping me and giving me guidance over the telephone, her opinion was always with me, as she lived during World War Two, she had experienced tough times, she had to cope with losing her first husband in the war, so knew what grief and fear felt like, I could relate to her and I did feel at ease talking to her. I hadn't spoken to her for a few weeks prior to seeing her on this occasion though, both mother thought a visit would be nice to cheer both of us up. It was great to see her as she had missed me and I felt really bad for not going round to see her, but what could I do? When you're in a mindset of fear, panic and anxiety, the outside world doesn't seem safe or acceptable. She inspired and still inspires me now. She lives life to the fullest and always enjoys what she does, no regrets or anything. She kept me busy and told me about all of her memories in hope that mine would seem fine. I did chores for her and it kept my mind busy. Of course I had many regrets and many rejected people, I didn't want to accompany myself with. Times were tough, but they are not always like that! You have to experience the tough times to get to the good times, she told me many of times. What great wisdom!

After that visit to my Great Nan's, I did go regularly with mother to hers. I spent time with her and enjoyed it. The time I spent with her got me out of the hell I was living in and allowed my mind to be occupied with others things like helping her do her chores, shopping, and gardening. I was able to talk to her and she spoke honestly and fairly which I liked. Lies shouldn't be told in families. I became friendly with some of her chums and I was known quite well around her area for being a loving, kind Great Grandson. I enjoyed being with her and certainly enjoyed seeing her and her chums contributing their efforts towards charity by taking part in a pancake run. It was a great time and great times were what I needed, the calm atmosphere was also great as I didn't feel alone or living in hell. I tried my best to put on an act and have my fake confidence up, I wasn't a very confident person, but I tried to pretend in some cases I was, whether it was wrong or not, I am not sure I can answer, all I know is I liked it.

When I left my Great Nan's though, I had to return home and it wasn't my house, not really. I didn't want to live in it. A home is where the heart is and my heart certainly wasn't in that dump, who's heart would be in a house they are scared of living in? Or kept awake every-night because of loud irritating behaviour? The only escape I had was to hide in my room with my iPod on and look through old photos of past times, some great. I couldn't change time and understood we must all move on, but going through photographs reminiscing about the old times, did help me, but I wanted to go back to live in them, the old times. They were better years of my life, which is the only reason why I can tell you that I kept trying to do this. The memories I had on that day when I returned home from hers were great. I just laughed all night and they were great citizens in the community. We need to respect our elders more, they've fought in our wars, kept our country safe and done their part for the tax system, so many of them are left alone and it shouldn't be like that! They need much more care and support, something that is much lacked unfortunately.

I didn't go on any more outings for a while; I only walked the dogs with mum and that was when she wasn't at work. Everyone else still was going out and enjoying life. I felt more positive about mine, going to school for my scheduled lessons did help, but I still felt as if everyone would be talking about me even if I/they proved they wasn't, again this was because of my anxiety and it does control how you think compared to how you used to think, it really does that, it can affect your whole mind. I wanted to go out but something was stopping me and it did feel hard for me to feel a part of the community as I hid away from it. I felt I was going insane and didn't know what would happen next. I had been suffering from depression and anxiety for around three years at this stage.

After a while of this feeling though, I decided that school and seeing relatives would help me get out of this vicious circle, I wanted to break out of it, but it was easier said than done. I had been encouraged to attend school, which Adrienne, mother and both of my counsellors did their best in doing, and I did attend more often in year 10. I didn't care for the negative teacher's comments. Whether that tutor I told you about, thought that she could stop me from attending by being so negative and rude I am unsure, but she wasn't even aware of how tough I had to be to not take the easy way out of life again, she was nothing compare to this strength I had, she was meaningless really. I had put up with a lot of stress and this wasn't going to break me. She was nothing compared to the problems I had put up with, spite is horrible but I did despise her, it was hard to show.

It was good to be back into my childhood routine, getting up early in the morning and getting ready for school as the adult does for work. I wanted this routine and craved to return to the school, it may not be my first school but it was still a school. I had people there I liked, people there I made friendships with, memories were made in that school, good and bad times. I wanted to make some more good times and enjoy my teenage years more than what I was. Schooldays are the best apparently, but in my case, they didn't really feel the best!

It did feel weird being back at school full time after such a long break away from it and just doing a few hours when needed to attend. It didn't feel as if I was there really. It felt like a bizarre dream. It was nice to return after my brief stint back at the exams. I was welcomed back by some more than others, but life isn't all about fan clubs and hate clubs. It's more about putting yourself first and not letting others bring you down. It was nice to see my friends again and even the negatives were not all that big. The teachers may not have been supportive but I got by with being me and the encouraging friends I had reunited with. I was glad that I chose to return although the coursework was as big as a mountain, it must have helped occupy my mind from constantly living in fear and worry, and most of all away from that dreaded area, I lived in.

The school counsellor was further happy that I was still attending despite doubting my wellbeing; I refused to acknowledge that I still had depression in school though, even if people knew, I continued ignoring it. I tried hiding that I still suffered from depression by telling all my friends that I was fine and that I didn't have to take the tablets anymore or attend counselling, both of which I did. My tablets were being given to me by my mother after all and she would notice if I hadn't taken them, before when I came off of them, I spat them out once she gave them to me, I would never recommend just dropping tablets as they do have side-effects on both your mind and body.

It was regretful hiding that I still had depression from them though as I was caught out. Another student within my year group was also attending counselling at The Newtown Clinic. They were not as humiliated as I was to see them at the clinic though. It felt like another knife being dug into my bad feeling back. I knew that after he saw me, that others may find out and I had told this white lie that I wasn't going to counselling anymore, when I was. At this appointment, where he saw me attending. I informed my counsellor that I was working of her strategy for me, I also told my counsellor about the good that was happening and that I found being back in school was helping me somewhat. I then told her about my two outings –into the town and to visit my Great Nan. She gave me so much encouragement and support and I will never ever forget that. I also told my counsellor that I was back and attending school. She seemed impressed with that but not so impressed with the whole "house horror" situation. The neighbours from hell and the youths made it difficult to discuss without feeling low and miserable. My happiness was always taken away by my worries and fears, it hurt so deep. The other situation was still feeling as if everybody was talking about me, the trust problem again. I told her, that I didn't show my feelings that well and that I needed to be hinted at sometimes to show them. I wasn't much of a supporter more of a worrier. She could help with the paranoia more than the neighbourhood problem as she couldn't fix that. Her advice was that we would have more sessions and continue reviewing my case and problems as well as come up with new ideas and techniques of which I was to follow. I only wish she could have been magic and turned the area back in a nice one and brought Melyssa back for me.

Think positively was the main key to getting out of my moods, see bright thoughts. I hadn't had any more violent outbursts at mum or dad, which was a positive, but I didn't have any arguments either, it all seemed quiet. I was able to see some good thoughts like the memories I had lived but not the good in everything. I had received a letter from a special friend. Josie, it was like she wanted to help even more than she did. She had sent me the letter to see how I was getting on and checking on my progression as she was always keen for me to achieve in life. She was one of the few people that really cared and offered support and that I could talk to. I was feeling touched after reading her heartfelt letter and my mood did get better after counselling, it had improved!

It did feel good for me to know that somebody else cared for me and put in a great deal of help and support as she was always encouraging me and giving my life's best advice. Between you and me, it more people took time out to help us (mental health sufferers) then we would probably feel more able to do things we once enjoyed, why should we be seen as different? We should just be seen as equals, even if we do need extra support and help.

The year, year ten, flew by at secondary school and it was the summer breaks already. I couldn't believe how quickly the time went. It only felt like a few weeks had gone by and

I did try my best to be in school every-day that year, even if I wasn't in 100%, I did try, more than I tried in Year 8 and 9. Time flies by so quickly when you are doing stuff and slowly when you're bored or not doing anything. I was dreading having 6 long weeks off from school though, everyone was looking forward to the break. I had nothing to look forward to really except more hell I expected. I couldn't understand why I had to have the break as they did so much to try and get me get back into school but the breaks were needed for everyone. I didn't realize how quick a year could go before.

Chapter 12- Year Ten Ends, More Trouble Begins

It was certainly unlike me, at the time before year ten to even not want a summer break, which was six weeks! I obviously did change that time, despite having my usual problems. Over the summer holidays though, it was one of the worst times ever, worse than ever though and I had been living with a great deal of negative and problems to the lead up of this. The area was still getting worse, so much worse infact! Even though it seemed as if it would get better with the dispersal order and increased police patrols, unfortunately for me and the other neighbours, that wasn't the case and more youths congregated and violence continued by the minute! It still wasn't an area I was happy to be seen in, let alone live it, those who have lived in a horrible area will not what I am describing, when your community becomes worried because of external people causing trouble, it really does take away the love for the area altogether. I had invited one friend from school round during these holidays and they never did want to come back again and they told me that as soon as they left. That felt awful to have that comment dropped like that. I still remained friends with them though as I could see their point as I was seeing exactly the same trouble. Their opinion just backed up mine that the area was terrible.

Loud parties, violence, smashed windows, noise and the police, all came in with the ridiculous summer break mix. I couldn't sleep again at this point so I was getting more agitated and less determined to fight but that extra strength lay within me and the people I had friendships with, even though I couldn't sleep very well previously, this was different, they were in the area the first thing in the morning and until gone midnight the following night, it was horrendous and certainly not something that someone who has depression/mental health needed! It was like every time something bad happened I could get comfort from knowing I had friends who supported me, but I did need a lot more than this comfort, I needed them eliminated, which never did happen sadly. It was an unforgettable feeling and one that I will never forget, when you build hate up so long for something, it felt like that, real hate for something I detested so much. It did bug me though and I wasn't as low as I previously had been, as I had been to school and tried to hide away from my problems, I didn't consider suicide either, which did show I had considered the last two times I tried and realized I wanted to live. Instead I wanted this crap to end and I wanted out of this area and for good. I couldn't have another argument with that neighbour or even with the youths, there were so many of them, or even start hanging out with them as I wasn't like them, I wanted to achieve and make a good life for myself, it wouldn't happen in this area, because I didn't go out alone, have confidence or feel happy.

I spoke to my parents and it was time we moved, it wasn't an area we wanted to live in any longer, the area had been cursed and nobody who lived in it could be happy. I had told both of my counsellors this, as well as Adrienne, the school counsellor. When she came out on a few occasions to do home visits to me, she even remarked on them, she couldn't understand why they were not at school either, but the fact is, so much more needs to be done to keep the young in school and training needs further investment, even for those who cannot be bothered, they should be made to be bothered.

I thought that we shouldn't have got this house, it was the biggest regret of all time, despite it seeming a nice area at first, the flat, our old home, had been there for my childhood years and I felt that was home, I felt I belonged there. It wasn't just brick and mortar, it was where a great deal of my life was spent and good memories had. I couldn't go back to that home no matter what though, but I could get a new home, a home that has our hearts in.

It was easy to say that I wanted to move but nobody would move into an area full of crime, terror and fear, would they? Fortunately though, the house eventually went and somebody invested in it, which was surprising really. And we got our newer, better house away from all the pain. I was finally free from it all, well the youths and awful neighbour! And was able to feel happy opening the front door to people, rather than looking around in vain, I could look out of my window onto a large park area, with countryside all around, not a youth insight! It was lovely and still is lovely, I know longer hear bottles being smashed or see broken glass windows, life did change for the better as soon as we moved. My family came round to visit and they didn't feel annoyed or hated visiting us, they felt more comfortable being at our house, because the noise from them simply wasn't around anymore, it was incredible. It was a great feeling to have people round visiting us again, even seeing my family who I thought didn't care was great as well, because deep down they must have cared, it was probably that they just had a funny way of showing it or didn't really understand, this is quite common with disabilities or mental health, the lack of understanding. Again, this is where the need to teach more about mental health and disabilities appears. I just spoke to them, the glimpse of hope was back in me, I didn't need that extra problem of 'living in hell' and I wasn't going to let it beat me.

No more loud parties to worry about or be deafened by, just peace and quiet in the countryside. Also the bullying and intimidation stopped, they know longer could try and wind me up, with their hurled abusive comments or trying to scare me, I never felt so good in that respect. Nobody should ever be bullied because they're different, once you've been bullied for years, like I was, you tend to realize that you didn't deserve it, you're not the person who asked to be bullied, but instead you were bullied because you were different and the people that did bully you are not worth being a part of your life and you realize that after time, they should never have been part of your life, but sometimes worry takes over, anxious feelings come up and that's what makes them part of your life.

Things had been finally been getting back into shape; I had left the hell neighbourhood, gained new friends in my new neighbours and enjoyed their company much. I was able to take my dogs out of walks again, go out and see new things and discover. It was something I had struggled with doing before, alone anyway. The atmosphere in the new area was better than the last. I didn't care what happened to that old area as long as I didn't reside in it, it didn't matter, we were finally free of that burden of worry for our area, because living in a village is nothing like living in a town, there is no comparison whatsoever. Good things do happen, even in tough times. I was glad to be able to feel the air again; I hadn't seen the outside fields and new people or anything special for years, but now I could as I was no longer cooped up it meant I could and mother and I loved it, as did father and my sister.

After getting settled into our new house, I visited my counsellor to let her know of the great news but my counsellor was sadly leaving the clinic and was happy to put my case to an open case, she was positive that I could get through my depression alone now and told me that, she said she has seen somebody who was filled of fear, turn into somebody who now appeared to not have so much fear, she wanted me to stay in contact with the clinic despite her not actually being present. I thanked her for all her work, effort and guidance in my case and wished her well. I was something I hadn't been for ages, I was happy knowing that I had one less fear to tackle. It felt so good to be happy, I could enjoy everything more than what I had done before and I didn't care if people were not happy for me to be happy as I had soul and my soul wasn't going to let anything or anyone defeat it not this time anyway, I had listened to judgment for most of my secondary school life, being told I would fail by staff at that school and being classed as 'mad' by some people, gave me motivation to prove them wrong, the only failure is those who do not try, I do try and wouldn't be this failure they portrayed me as. I would miss my counsellor but things must have been getting better for me and she probably knew that moving house would have a lot of good for me. Moving was one of the best things I ever did, after moving from the flat, I felt trapped as it had been our home for years, but our new house showed that not everything is bad in life and even with mental health, factors can help us feel up-lifted and happy.

Shortly after moving into our new house, a new home card was sent by Josie; I rang her up and thanked her for it. She was so pleased that we finally sold up and moved into an area we actually felt happy in. She congratulated me and wished me every bit of happiness for the future. I was grateful for her support as she was one of the few that helped me through the tough hard times I suffered and that will never be forgotten. My Great Nan also was glad that I was happier and away from all the trouble and bother. They both were great. It showed me how much some people go out of their way to help others and I think that their support along with my parents and 'trustworthy' friends saw me through the struggle living where I did. I just regret not being able to move sooner. Nobody deserves to live like that. The new house was a great way to show positivity, we have a large garden and I started gardening, some people may say this is for the elder generation, but I can happily say that gardening helps me now, on my tough days with depression, I enjoy gardening and it calms me down. I recommend you to start gardening, being out in your garden in fresh air does help you, as well as burn calories and provide exercise. This was my first hobby, since losing my old ones when living in the other house, battling with depression. The garden soon was looking loved and is something I enjoy doing.

Sadly after moving house, the following year Josie died. It was very painful to hear of her loss, I couldn't believe it and I regretted not spending more time with her. It was awful; I never got a chance to tell her how much she meant to me or how much help she had given me. I think she knew deep-down how much I valued her and held her close to my heart. I told her on the last day of primary school, before waving her goodbye, but she didn't think it was a big deal, that's what I admired about her, she didn't want anything back in return, and she gave… something many do not. I was lucky to have known her and even now I still have flashbacks of the good old schools day I had with her. Her helping me and others was what she did best. I did class her more of a Grandmother, than a friend, she always supported me and I have assurance that she is around now, especially as I keep having these flashbacks, I only regret not being able to see her in person. I never attended her funeral which was a devastating moment as she really meant a lot to me. I didn't get told about her funeral, I found out she had passed away when I visited her house and a neighbour told me. It does feel weird without having her to write or talk to but she is always in my heart and lives on in memory.

She was a person who cared and worked hard with others, she wasn't selfish at all and she will be loved and missed always. She never had a bad bone in her body. Josie knew about what I wanted from life like my law career and my business studies. She encouraged me to go for it and I just started college when she died, so sadly she didn't see me get through the subject. I regret not having more time for her as she deserved a lot more time than I gave to her. I hope she is looking down on me and is proud. It is very hard to lose someone without telling them what they meant to you and I didn't even say goodbye to her either. That's the worst, horrible feeling eternally. I guess that sometimes we never do get the chance to say goodbye or thank the person for what they have done for us, which is why I was distraught.

Yes, you may have seen I mentioned I started college. This happened a year on after living in our new house, after that horrible summer holiday at secondary school, I didn't go back to secondary school, which meant I didn't sit my GCSE exams and left school with no qualifications, they may have thought I failed, but in actual fact I didn't, you'll see why soon. Josie knew all about my plans to apply for college, even though I worried I wouldn't be accepted due to not having any qualifications.

A few weeks after her funeral and being told she passed by. I visited her grave and told her how much she meant to me, despite her not being there in person, I could feel her presence. It was as if she was standing next to me with her hand on my shoulder and listening to me, I started crying as I really couldn't believe she was gone, her picture was on her tombstone and it felt wrong seeing her there. I wanted her back, but I believe she is still here and I like to believe she is always around, it does comfort me.

Chapter 13- Good Happens, Life Changes for the better.

Since moving house I had been planning a future for myself, some may say because I didn't have any GCSES due to not attending the last and most important year in school, that I wouldn't have much of a future, the school staff had told me this long enough, but this wasn't the case. I had applied at a local college, applied for work and even got more plans together to start seeing friends. Despite the move, I did still have my depression, because it wasn't just the old house that caused my depression. My anti-depressants were still being taken and I still had my sleeping tablets but they were what helped me I guess, even if I couldn't see that, the doctor wouldn't have given to me for no reason.

Within a few weeks, I had a letter from the college and a welcome pack; I had got into Huntingdonshire Regional College and was so thrilled with the result. I feared not getting in as I didn't attend school all the time or have any qualifications, but I got in. I was over the moon with the news, I so looked forward to going to college and meeting new people and studying something which I wanted to do, I had chosen to do. It was great to know that despite everything, I could continue with my studies. So I started preparing to go to college, I had none of the 'negative' feelings which I had after leaving primary school going into secondary school, the only feeling I had was 'apprehension' over what the college and class would be like, but then everybody else would have these feelings too, I expected.

My first day at college was finally here, it did feet weird and very odd. It was very different compared to school as the atmosphere and buildings were completely changed. College was poles apart to school, college was a more grown up environment and less people backstab you or betray your trust. The first day of college was challenging though, as I didn't know anyone but I had the encouragement and was out to prove everyone wrong about me, especially those who said I would fail. I never understand those, especially teachers, why they have to try to belittle students, especially those with mental health. Could they not see I wasn't feeling full of motivation, happiness or life? It makes me sick. I may not have been able to go to school all the time but college was going to be different. I was a different person since moving, a stronger person. I was determined to make it to the end of this first year at college and I didn't care who tried to get in my way as long as I achieved. I just wanted a career and I made sure that I worked hard to get one. Part time work didn't really go well as the jobs were never really for me, I wanted my legal profession but without the qualifications I couldn't have one, which is what made me more determined to continue studying. A degree is needed to become a lawyer after all, so I decided that I would stand to become a student representative, which I won and would help the class.

I met some people who eventually turned out to be good friends at the time- Adele, Ghulam and Theresa. We all met on the first day and it was great to meet them and go on the year journey with them also. They brought happiness and good memories for me. Adele had previously studied at the college, she was quite well-known around the college for her pranks and mischievous behaviour, Ghulam, he had never studied at the college before, but was a really nice, down to earth man who I never have seen since that year ending, and Theresa was nice, she came from Papa De Guinea and had a goal to become a child minder. The tutors were also great and very supportive- Catherine and Belle made me feel so welcome at college and they really went above and beyond to make sure that I and everyone else achieved and I really respected them for that. Catherine in particular, always had time for me to chat with. She would encourage I talk to her about anything, so on my bad days, I did talk to her, I didn't want to keep my worries hidden anymore, and after the first week, I believe I was just worried about the work, but I needn't have been, because I passed that course. I remember we had some great cooking lessons as a class and hearing talks about life within this group.

It was a great group and I didn't have anything to worry about really, it wasn't like my school experience with nasty tutors or bullies, it was more of a 'you want to be there' environment and that makes the different I feel. Both tutors gave their best and brought the subjects alive. They had respect for the students because we had respect for them, that's also the difference. The learning environment was totally changed and more welcoming. It was a lot better and happier than school, I didn't have loads of coursework from different subjects adding onto me or the pressure I was already facing at the time, I also didn't have some awful teacher who only wanted gossip causing my life hell, I had pleasure! College is a place I wanted to be and live the moment for. The friends I made, only stayed with me for that year though, as we all went our separate ways and unfortunately, Ghulam didn't make it on the course he wanted to do at college- Business Studies. It did give me confidence though and my worries about starting this college that most probably have was only that, worries. It never did really become a fear, because after going for the first week, I would have confronted it, if it felt as if it was a fear.

Sadly again that year went by quick, it really did go by quick! Before the year was finished though, we had to consider our options, what with us studying a Pathway to Employability Course, I was considering my options and thought about getting a full time job but full time jobs were hard to come by as the climate was facing financial uncertainty and was uncontrollable and I hardly had the relevant qualifications for any job I considered appropriate anyway. So I decided to apply for a business course within the same college, I had no knowledge of business and didn't even have any idea on how a business was run. Ghulam and I both applied for the BTEC First Diploma in Business, and as mentioned he didn't get in. I thought, oh no I wouldn't get a place. I wasn't expecting to get a place on the business course so to my amazement after my interview with the course leader, Greg, I was surprised to learn that I had been offered a place. I was so happy and I knew Josie would be too. I told my mum and she was also full of enthusiasm. It was nice to tell my family, that my hard-year of studying did help me open another door and I was going forward with my life and not back this time, not wanting to be in the past, but in fact the present. Greg must have seen that I had passion and wanted to learn, that is why I believe I was offered a place on his course.

The long summer break came up that year and I was glad that it came up! I wanted the rest of the enjoyment from being off from college despite missing a lot of new friends within the group. None of them returned to the college, it was a shame, but then that's life, everyone goes their separate ways in the end. I did miss not seeing them around the summer holidays.

The interview was a great one and I really felt that the subject- business was for me, the tutor Greg was a man of many different talents, he appeared to be very knowledgeable, honest and factual not like some 'acting' teachers that only tell you what you want to hear. I had been speaking to other students who had done business before and they recommended it to me and even said that

Greg was the best teacher they had ever had. It made that course sound like my best decision so I accepted my place and told him in person. I would miss Catherine and Belle though, as they were my first tutors, but I would still see them around the college after-all, which was good.

I also applied for a new job in the break, which I got. I didn't really want part time work but I was going to give the job a chance. I was a Customer Service Assistant within a large family run card shop, Clinton Cards infact. I was grateful for the chance. I really hadn't worked properly in my teenage years as I wouldn't have been able to work, so this was an experience I won't forget. I was working within a team and working for the customers. I always remember wanting to get some experience of this whilst I studied business, as it gave me the experience and real life knowledge of the retail world. The job hid the pain of losing Josie and of my depression, as I was pre-occupied with meeting new people and working. I enjoyed working for the company and even met some great people, some not so great too, but that's like everything isn't it? I had celebrated and worked through the seasons- Valentines, Easter, Halloween and Christmas with that company but after being with the company for a year, my college studies forced me to resign.

The workload was far too big to be able to do both and it wasn't fair on me to rush my assignments due to not having time and I didn't want another worry of not having them completed on my stack. I really miss that job and most of the people. It was a decision that I made that sometimes I wish I could change because I really liked the job however I guess I made the right decision as I always put college first not money. College and studying eventually will help me into my law career.

The break went by quick and I started getting prepared for college. I didn't know anyone on the new course so it was worrying but I had been to the college so I knew what that was like. I also didn't know anybody on the Pathway to Employability course I just finished studying, so that felt normal to me. I felt more positive and did have to take any tablets still at this point, I still take them now, and they help. I was still coping with my tablets which felt great after being on them for such a long time; it was a relief that they do help and don't go against the person that takes them. The new area, the new start and my new college all helped me be happier and enjoy my life more than what I would have done, especially if I was at that school and old house.

The business course was great; we were all made so welcome by Greg, our course leader and everyone was so friendly and really made it a year to remember, we all got along well. We had our induction days for a week, I sat next to a person who I instantly liked his name was Omar, I had seen Omar around the college on my first year, but never introduced myself to him. Omar seemed very logical, especially in the sense of Computers and helpful. He was a person that I could relate things to as he seemed to have a good understanding of depression and I offered my help to him with my advice even if some of it was unsolicited I am sure he appreciated it. I was glad to have met a new friend on the first day although I wasn't all that worried about not having friends; it is nice to have them. Omar admitted to me that he had been bullied before, so I knew we had this in common. He had studied computer studies before the business course, so knew the college as well as I did. I then met a few other friends, Sam. P, Sam. K, Will.J, Jake. C and Gary. F. Shane, one of my closest friends also was introduced to me later, as he joined the course a few days later.

It was also so nice to see how much things had changed; it felt like the goods things I had been waiting for had finally arrived. I still had my confidence issues though and they may have caught me out a few times but I mean the fake confidence was there for me when I was fighting depression and still is there for me now, I guess that it is there for a reason although I am unsure why, it helps me get through things.

When I lost everything, well felt like I had lost everything, all of my happiness had gone, this fake confidence would help in some-ways. The first week of the business course seemed great, everyone befriended each other and I had a group of friends again, just like at primary school, but only in the present time and not the past time. Within the next few I got loads of assignments and I was determined that I would try and do them all well before the submissions date and I did. A new member joined the class called Shane and he seemed very opinionated and I liked that especially as I was like that before the depression had affected me. Omar and I introduced ourselves to him and the group was formed along with another new lad called Sam. It was a great group and I even had another group of people that wanted me in their group called Ryan and Yasmin. They were both also nice people who I could trust and they joined in with Omar's group and we all got along fine until they left and it caused problems within the group, but luckily they got sorted out by good communication and teamwork, mainly on their parts, because I tried to avoid any more arguments, at all costs. Again, on this course I was a student representative, I enjoyed supporting the class and they must have enjoyed having me support them and their ideas.

Other than us losing our classroom and being moved to the other side of the college, the year went really well. Everyone helped each other and the tutor again was a legend, he really was. A determined and strong man who stood up for the group and worked hard to make sure all achieved. I achieved a distinction plus in my BTEC, so I was the top achiever in the class. I did put in time to help the other guys out though, because by the end of the year, Yasmin unfortunately left. I made sure I helped them the best I could, so they got the grades they were happy with. I was very happy with my achievement and the progression I had made. Before I started the course, I had no business knowledge and then I put the effort in to learn and I was the top achiever. I now had business knowledge which I never had before and my older teenage years felt achievable.

After studying the BTEC First Diploma in Business, I decided that I wanted to study the BTEC National Extended Diploma in Business, which unlike the previous business course, was 2 years and not 1. I had an interview with a finance tutor, who was amazingly knowledgeable and supportive, called Vanessa. She had to take my interview this time, because I already knew Greg. She asked me what I had achieved and I told her, I had achieved my Pathway to Employability qualification, as well as my BTEC First Diploma in Business, at distinction plus grade. She seemed impressed with this and then asked me a few other questions, which I answered honestly. A few days after this interview, a letter came to my address from HRC to inform me that I had been accepted onto this course. It felt great. I knew that Sam. P, Shane and Will. J would also be applying for this course, so this time I actually would know people on my course, Omar unfortunately wouldn't be joining us as he was doing I.C.T at the same level but different course. It did take some getting used to, not having Omar around on a daily basis to chat with and Jake and Gary left college that year. Both of them were good friends, they had laughs and jokes, some I didn't understand because I have a little sense of humor, but I enjoyed their company and presence in the group.

That year came by quick, in 2011 I started my next business course and to my surprise Sam. K returned. He was in our old group, but told us he wouldn't be coming back, but he did, which was good because I had another person I knew in the group. Sam. P, Will. J and Shane all were made offers onto this course too, which they accepted. It really was great to be back at college and I enjoyed it. I became good friends with my tutor Greg and we had three other tutors- Vanessa for finance, Ken for business management/environment units and Alison for our team-working/communication units. I had Alison the previous year for four modules, so it was good to have her again, Ken and Vanessa were both full of knowledge, just like Greg and Ken had so much knowledge, it has hard keeping up sometimes! You had to write everything down, he could write many books of wisdom with his invaluable experiences, he really could. Vanessa encouraged me to get a distinction in the finance module, even though I hated Maths and failed miserably at it, she was by my-side making sure I gave it my all. I eventually got a distinction, but her advice of don't treat finance like Maths, is true, because it is so different yet similar in some respect. She made time for me, which the school tutors didn't and that's what counted for me success in that module. I joined the college council this year again, there was only 4/5 of us on it, but we tried supporting the college, we loved, in the best way we could.

Not everybody saw eye-to-eye on the first year of the business course, we had rows, disagreements and I had many rows, because I couldn't express my thoughts correctly, I had to tell them to my tutor so he could sometimes, it all become a bit of a heated row sometimes.

Fortunately, we all got on in the second year and final year for me at college. The whole class were friends and had a great time together, I got on with the people I had always been with and kept in touch with those who left, as well as Omar, often seeing him around the college. Ken offered me a chance to study a Chartered Management Institute (CMI) qualification, as he saw how well I was doing in my Business studies, I accepted this and along with the other Matt in my business class, Shane joined me and we ended up studying the junior management qualification every Tuesday evening for ten weeks, that went by so quickly too and I kept on track with the work and had little worry. I was keeping busy and that helped my depression, I trusted the people close to me in the class and that helped me, although depression doesn't just go away, on a good day, you do feel really positive and vice versa, on a bad day you can feel really negative.

On the first year of my two year business course, I did 9 different units and achieved distinction grade on all nine of them! It was amazing, having studied the subject for a year before that, I built knowledge and I felt so satisfied with myself. My college principal Susanne, had opened a competition to enter to open the new college building and she chose me, Anna and Liam out of all of the applications to do so. Below is a picture, which was taken by The News and Crier. Left to right- Liam, Susanne (Principal) Anna and me.

This felt a great moment, I had been noticed by the principal and my desire to help the college, as much as they had helped me must have been showed.

On the second year of college, I decided I wanted to give something back, so I built my own website beatdepressiontogether.webeden.co.uk . This website was designed as I felt that due to all of the government cutbacks and waiting lists for counselling, there is a need for support for those with mental health, I have that and know what it is like. I've lived with it for many years I thought, so why not. The website has come together and I really am proud and happy of it. For my website, I was nominated and become a finalist in YOPEY (Young People Of The Year) competition 2013 and won Student of the Year from the college, as well as a Community Inspiration Awards from Luminus. I never have received or been nominated for so many awards and it felt great. I studied 12 units this time, making the total of units 21, the amount of units was 18, so I did three more, because studying helps take my mind away which leads to my depression being occupied, I hope that makes sense. I achieved distinction on all of my units, which meant I had 21 units at distinction. I have never been so happier, my overall grade in my BTEC was D*D*D*, triple distinction, plus, plus, plus. I was so grateful and knew I chose the right college, even if we did have our disagreement on this year over the college making cutbacks and cutting back a trained counselling lady. I was glad I chose HRC, because I felt valued there and wanted.

To add to my qualifications and achievements, I was chosen by a board of employers to receive an award, which a friend of mine named Fiona, encouraged me to enter. I pitched my website to them and was chosen the overall winner, it felt fantastic. So much had changed for me and I was no longer seen as a failure in my family or friends, my trying spirit came alive and I tried my best to give everything I had, this shows that just because you have depression, you shouldn't be seen as a failure! You should be understood properly. I helped almost everybody on that course with their coursework, understanding or application of knowledge, because I built my knowledge of the many different areas in Business up and even surprised myself, business was definitely the pathway for me. I even had studied another qualification CMI at the same time and passed that too.

Above CMI graduation, Left to right, Sam K, Shane, Robbie and Me

For the Luminus award, a lovely, kind lady named Bridget, nominated me for it. She visited me at college after hearing about me and shared her concerns with me. We both agreed that there needs to be more understanding and support, so much agreeing occurred. I gave her a copy of my book which she shared with her crisis ward and it gave them hope and encouragement, she really was lovely and must have thought highly of me to nominate me, I am truly thankful for her and hope she long continues to support mental health.

I was happy with my achievements, friendships and new memories which I had made and it sure proved to people that Matt was back and I will not be going anywhere and despite still having depression and anxiety, I don't think it will ever leave me, not properly and having a few bad days, I think I have learnt to live with depression for good. I still take my tablets for support and am glad I didn't let the negative people win as they really are not worth it and they should all grow up and look in the mirror before they judge!

Epilogue

Having Anxiety and Depression is the biggest challenge that I have had to face; it was very hard and emotional. Some days didn't feel like life was worth living and other days I felt okay but not very happy. Sometimes it is still like this for me, but I do have many happier memories now, some are not so happy, but that is part of life. It was like a rollercoaster a never ending one. I appreciate all the help and care shown towards me during my tough time and urge anyone who has any problems or suffers from depression/mental health, don't hide away and to get the help they deserve and need. Don't be afraid or frightened getting counselling as it is the best thing to do, it helps to support a person that has gone through the condition and it helped me quite a lot, and I feared it to begin with! This shows how things change.

You may feel some days are worthless or difficult and you may even feel like giving up but don't ever give up, there were many times I felt like quitting and giving up, but I didn't, something kept me strong, whether it was support or my trying spirit, I just cannot answer. One thing that depression does is show your strength and whether you're capable of fighting it, I say fighting, but it's more of a case of living with it that counts. Sometimes it takes years for depression to be fully defeated and some people have depression longer than others, but it doesn't mean that you are different from anyone else it means that you have been set a challenge. The challenge for you to win, I am sure that everyone who suffers from depression is strong enough to fight it and especially as I was a teenager when I got it, it made me reassess my future and I am actually glad I have depression even if it was a problem, it made me realize that everything I did or everything I didn't do happened for a reason. I don't believe depression always leaves every sufferer who has it.

Life is worth living and I am extremely happy that I am still living mine and the negativity has disappeared, well most of it, and it really does show that you can win even if you don't think you can or it seems everyone is against. You must try to live with it, never consider suicide or the easy way out, I regret my attempts, try talking to somebody close to you.

Depression is a subject that many people must go through or have been through in their life time, it wasn't just me that went through it and sometimes you never even know you have it which makes you think it may just be change. Understanding of it needs to be increased by outsiders though and in schools.

Depression can appear at any age time not only just in the teenage years, it can happen when a baby is born also known as postnatal depression, it can affect adults, when a family member passes and especially when the hormones change as this affects the way we see things in an everyday point. It can also affect the elderly, they also need more support! Depression needs support.

Sometimes fighting depression can be hard and even harder if you do not know you suffer from it, it is curable but you have to know you have it first to fight it, the cure is being able to live with it. Do not go and try to fight depression alone, I would have never won if I did, even though I kept trying to tell myself I would have at the time. I had help from professional counsellors, close friends and family members without them I probably wouldn't be here now, so make sure you have support and help even if it isn't with family, the professionals always help and do their best to make sure the service they are offering really works. I have anti-depressants prescribed to me every week by my doctor and it isn't a bad thing to take these as the tablets help with the 'low feeling and bad moods'. Everyone with depression will vary with different cases, so my cases of fear and hatred probably will not be the same as yours but all depression is related.

I wasn't aware of my depression at first, remember that sometimes we refuse to accept change in our lives and I was rejecting my change, I didn't want to move on from a stage in my life that I simply enjoyed and I was fearing quite a lot which I didn't really have too, life has a funny way of sorting problems out. It isn't accepting the change that is the hardest thing; it is accepting going through the change. At the time I was a young child who simply was confused about why we had to go through the primary-secondary transition, it wasn't something you do every day and it broke my routine and changed me. The tough times after, do not happen to everyone with mental health, I accept that, but it was my journey.

Do not let people's opinions affect your mental health, some people just shouldn't judge and I was judged quite a lot by bullies and even some tutors at the time of my secondary school time. I was allowing them to get to me and take control of my mind. I was under a mind game of paranoia, anxiety and fear. I hardly ever feared before depression and I was going through a mixture of emotions as well. I allowed their opinions to get to me for so long and then after so long of their hurtful bullying and negative comments, I just realized that their comments were worthless, I had become ignorant to them. Why should I care about their comments? And that's what you always need to have in your head, if you need counselling- accept it; you're getting more help than the bullies! Counselling isn't a bad thing, most people seek some form of counselling in their lives and the counselling improves and makes you feel a more understood/stronger person, so it isn't a place where you go if you are mad like some people suggest. I was told I was mad attending counselling, as I shared, but I wasn't mad and the people that think counselling is only for 'mad' people obviously have no IQ! Never fear counselling or getting that extra support!

I was always told to think outside the box, I might be having counselling now but think of the end result. The end result of counselling would be that I would have the extra help and plans and to put the plans into motion to help my health and my way of thinking. I really appreciated all the help I received from the counselling team as they were great and very knowledgeable. The end result for me was being able to see my life more positive and have more will power. The will power helped me to stand up for what I was, what was right and also to fight the negatives more along with the bullies.

Again the counselling needed between individuals varies and your counselling sessions may be different to mine, you may need more sessions or less, it doesn't matter all that matters is your health, that's a priority.

I remember my first counselling session; I didn't want to go and was worried. I didn't think an outsider would be able to read me. I had many thoughts going on in my head, and I was worrying. An outsider didn't know me, what happened if they judged me? They might have even laughed at my depression. But they didn't moan, laugh or judge me. They helped me. They had the knowledge and support I really needed. It was nothing like I thought.

After reaching out and accepting help from the outsider, I felt I was changing again. The pills helped me but it wasn't just them which helped. It was my views of life. After the suicide attempts, I was encouraged to attend more sessions and the school didn't even seem to care or help me through any of it. I felt as if they had little idea of how depression sufferers cope, they didn't think that I had it, they just thought I was acting up. It was really annoying to think of how poor the school was, I had picked the wrong school to attend! My attempted suicide was built up of all my fears and worries and I regret even trying to attempt suicide! Do not try it, think of your family that you will leave behind, it is not fair and you are a better person than that. You can get the support, I eventually realized I had.

The pills helped, my counsellor helped and my family. I was getting help and accepting it which made things easier. I was able to develop better confidence skills and became more able to fight my suicide thoughts. I had cards which I looked at to make sure I was getting support at home and wherever I was. I just turned a feeling card over and saw what I should do if I was in the situation/ feeling. I found the cards a good idea and found the cards to help, perhaps having something or someone showing you the way, isn't such a bad idea?

Remember the difference between good and bad. Good isn't often seen or appreciated and when you do something bad, that is always seen and never dropped. Being bad and seeing negativity all the time isn't right, people who judge a depression suffer for being negative isn't fair or right on that sufferer. I found it hard to block out some of their comments at first but then I was finally able to shut them out after getting the counselling and support on how to do so. However, some people may suffer in silence because of the petty and unneeded comments, which infuriates me. ☹

I am actually glad to have depression. I know that sounds a weird thing to say but I was able to develop myself into a better person. Depression made me see everything differently. The fearing, the paranoia and the hatred all had to be understood. I also had to accept that change happens for a reason and some situations and events I just cannot control. The depression may still be with me, but I could still try and that is exactly what I did. After a while, without even knowing about it, I wasn't speaking critical about myself, family members or others, I wasn't being their biggest critic or being so negative about them. I was speaking to them rather than feeling against them, like I thought they were against me. It was well worth the effort being able to get back in contact with family members that I had disagreements with; I just think some people never truly understand depression at all and they could do with the help and knowledge before thinking that we are only out to cause trouble as that is totally untrue.

The fear never did really go away, I always think of things before they happen, still now I think over things 5-6 times before I do them as I don't like jumping to conclusions and regretting things I do. I was full of regret over the changes that had gone on in my teenage years but they were out of my control, so the regretting shouldn't have been worried over but that's what it was like for me when I suffered. And the paranoia was ruining the only relationships I had with family and friends as I misjudged all of their actions and commitments towards me. I couldn't control the paranoia and that was the worst thing to cope with. I had no confidence whilst I suffered or self-esteem and I wouldn't stand up for myself either. I was becoming a shy, silent person and sometimes I would be an outspoken, loud person. I had two personalities, two completely different personalities which made living my life very complicated and somewhat tricky.

The fear of losing everything I was, made it hard for me to accept the reality of a lot of changes that I went through, this did make it harder. I got taller and wider, my voice became squeaky, acne appeared, and I was changing in body as well as mind. You only ever experience this once, so not experiencing the change before made it very confusing and mystifying. I was being challenged by these changes and the biggest achievement of all is accepting the change, which I didn't do straight away but I eventually did. Puberty has to happen after all!

Nothing I did alone worked, I tried ignoring the actions of myself but they would soon come back to me. I thought I couldn't do anything right. I was always told to stand up for myself and stand up for what is right and I did as a child but that disappeared when I became a teenager as I was vulnerable and edgy. Sometimes when I was quiet I think my family appreciated the peace but when I was loud and outspoken I think that caused big annoyance to them and brought us a lot of disagreements. I was fortunate to have family who I could take my frustrations out on though even if they all didn't help much, they were still there for me in their own little ways. I hated most of them whilst suffering, the hating people thing seemed normal to me and it seemed right. I never hate anyone now though as the word 'hate' is very strong.

I luckily had Josie, my parents, close friends and Great Nan who all helped and supported me most. I still really miss Josie now significantly, but am glad I had the chance to know her, she is with me right now and even without her words of guidance, I now have my own, partly thanks to her. ☺

As I couldn't see the good in my family, I thought they were like the other critics I had. They must have hated me, they must have wanted me out of the way I remember thinking. They must have seen me as a big thunder cloud coming their way. They probably would have found it easier without my sharp tongue attacking them. It wasn't me though and I was able to see that, it wasn't my normality even if they couldn't. They are still here for me now though, which shows they do care.

Teenagers have so much bad reputation and it isn't really fair on them. We all have to grow up to become adults and being a teenager isn't a bad thing at all, just because we have confusion and worry over different things like relationships, families, money and schooling it doesn't make us any worse than adults. I think that teenagers should be accepted more into society rather than being seen as a big problem, mainly due to youths and gangs. Perhaps if they all get the same help and support from a few family members as I did they would be happier. The bad reputation I had however, didn't really bother me in the end, my ears were burning all the time and I knew they were talking about me, yet I had listened to it constantly so was able to switch off.

I found not doing anything also made my mood worse, for example the 6 week holiday after leaving my primary school. I had a very long time to sit at home and worry about the worse that could happen to me and I was coming up with all sorts of problems that could go on and happen. That is when the fearing started, I think looking back at that now that I should have made it my prerogative to have other things to do, like I do gardening now. Perhaps that would have helped me? I wouldn't have gone through half of my worries if I had other things on my mind. But I cannot change that now and I just look and think that it all happened for a reason. I advise you to get different things to do though.

Teenagers need things to stimulate them and make them happy. If they are not supplied with knowledge or help they will look for it. I know that some end up being complicated but is that because they are worried and fearing? They shouldn't be seen as a complication like I was seen as; they should be seen as an objective and priority. Perhaps if they had more things to do and social clubs to attend then they wouldn't be thinking things over or causing problems. I wish I had that option to go to more clubs and socialize as I thought that I couldn't be a friend to anyone after the problems I went through with my last friend group, sitting in silence and worrying wasn't healthy.

I should have talked it through long before I did, I should have told my parents my worries.

Confiding in people is really helpful, a problem shared is a problem halved and I needed my problems halving before everything got to me. Unfortunately I didn't confide in anyone other than Josie and my parents, but I didn't want to keep bothering her with all of my problems. If I had more people at the time which I could confide in then this could have helped me further, confiding in people isn't a bad thing at all, as-long as they are genuine friends... It doesn't make you weak or a coward, it helps you share your problems and face them like a big person. Having the confidence to confide in people though is the key, if you are not a confident person like I wasn't at the time, then it makes things a lot more complicated. Shy people tend to stay quiet but they shouldn't be encouraged not to talk to people or not be accepted because of their personality.

The confidence of a person does take a few kicks before you see the real person deep within. Confiding is a big thing for some people and they shouldn't be mocked, just heard. Talk to your closest friend or your family. Even if you cannot trust them, talk to somebody.

It is true that people with depression do feel low or uncared for and this is the part where having close family and friends around you helps! Families should always stand up for each other and be close, you shouldn't be worried about telling your family anything and you should be able to trust your family totally. If they all totally believe that then they will support you in your time of need. We all have hours of needs. They will support you with your depression and help you through the tough times ahead that you may face. Families are rocks of life.

<u>My greatest advice for Mental Health suffers would be:</u>

- Do not go it alone, even if you feel you can, you must accept help and the professionals may not know you but they will help you. Even if not a professional, a friend or family member could help.

- Don't let the unkindness of others beat you, you can be a better person for not retaliating to their criticism of you and ignoring it. Ignoring it may be hard, but you will get used to fighting the bad things and realizing what is needed to help you cope with depression.

- Don't change who you are deep within, depression and anxiety are not a bad thing that is given to you. It's a challenge. The depression challenges you on so many different levels and on so many things. I was challenged over fear, loss, confidence and lack of trust but eventually I had fought these, one by one and won, sometimes it takes time. And every-day isn't the same with depression, so even if you do notice change, it isn't always bad.

- Don't worry until it happens. We could all sit and worry about the future and what may happen to us when we are in it, but until it happens it really isn't worth worrying about, it can be hard though, I've done it all so many times. We all may fear some change but it happens for a reason and the change needs to be adapted around our current lifestyles, if we are going to stand a chance with the new changes. It is like going to work and you have a new manager, you either get along with them or you don't, it's the change that makes things differ and sets upon more challenges.

- Never, ever take your own life. I tried and regretted it. You will be letting others defeat you, especially if you suffer in silence. Make sure you stay strong even if you feel you have lost every part of you, you can be strong! You can always visit your G.P, speak to somebody or visit my website and contact me for support.

- Ignore people who don't matter, everyone has an opinion and some will be more offensive/harming than others, what is the point of worrying about somebody's opinion whom you do not even know or care for? Like I did about that tutor, she was worthless to me, but I let her bother me for some-time, until realizing she wasn't important. That's when you begin to feel that they are irrelevant and that helps with depression.

- Don't hide away from your problems, face them! Hiding will only make the matter worse and your brain will go into overtime with worry and with loads of 'what if's'. Stop and think about the problem, is the problem really worth worrying about? If it is worth worrying over, what can you do to fix it? There are always solutions even if they are not that easy to see, if you think over and over again, you will see them, but don't keep concerns to yourself, share them if you feel able.

  And finally, never let people see you as a failure, since leaving secondary school, I have proved so many people wrong about me, I have academic success at distinction grades, gained entrance to university and have received many awards, so stuff those who doubt you and care more about those who have your back and support you. I support you all!

I hope that you have gained a better understanding of what it is like to go through or have depression and anxiety, and how I have changed and fought depression, I have fought it by learning to live with it, but after reading about my story with it, you can see this hasn't always been the case. It can be an emotional rollercoaster ride but with the right support and guidance you can fight it like I did. It will not always be easy and not everyone will understand depression but one day hopefully everyone will understand it and understand the people who are going through it better. The more someone gets noticed, the more the person feels appreciated. The better the understanding of them, also helps. Understanding doesn't make them feel isolated or alone and having depression can often do that. Let's beat depression together!

Best wishes for your futures, remember together we can Beat Depression and increase the awareness of Mental Health! I hope you've enjoyed reading my story and found inspiration for your journey in mine. Even when times get tough and you feel isolated, remember there is always some support or good around you, I now see that.

Stay strong, get it touch and thanks for purchasing my book.

Matthew W.N Clifton

Please check my website out for further support:
www.beatdepressiontogether.webeden.co.uk